Groups and Markets

Hans Gersbach · Hans Haller

Groups and Markets

General Equilibrium with Multi-member Households

Springer

Hans Gersbach
Department of Management, Technology and Economics
ETH Zürich
Zürich
Switzerland

Hans Haller
Department of Economics
Virginia Polytechnic Institute and State University
Blacksburg, VA
USA

ISBN 978-3-319-60515-9 ISBN 978-3-319-60516-6 (eBook)
DOI 10.1007/978-3-319-60516-6

Library of Congress Control Number: 2017943239

Printed on acid-free paper

This Springer imprint is published by Springer Nature
The registered company is Springer International Publishing AG
The registered company address is: Gewerbestrasse 11, 6330 Cham, Switzerland

Preface

The book reflects on our collaborative efforts since 1996. We believe that households, their composition, decisions, and behavior within a competitive market economy deserve thorough examination. We entered unknown territory when we began our investigation of general equilibrium models with multi-member households. It turned out that we entered very fertile ground. We became increasingly convinced that general equilibrium analysis can still provide novel and relevant insights into the workings of interdependent economic institutions even though the bulk of economic research has shifted toward other areas. The focus of our analysis lies on the interaction between households (and groups in general) and competitive markets for commodities. We revisit, organize, and reinterpret material from prior publications. But we also include new material and perspectives: Some of the relations to the literature found in Chap. 10 would be difficult to develop in a technical research paper. The suggested policy applications in Chap. 15 collect ideas forwarded in earlier publications but new ones as well. Some of these applications are straightforward in view of our theoretical findings while others require further thought. Two chapters contain entirely new and unpublished material, Chap. 5 that includes production and Chap. 12 that is devoted to the risk-sharing capacities of households versus those of markets. Clearly, Chap. 12 is only a beginning. So is Chap. 14 that merely delineates the confines of a model with public choice of public goods and their funding. Other topics are barely covered as well such as household production and taxation. Obviously, the research agenda remains open-ended.

The chapters of the book are grouped into four parts.

Part I: The Theme. That part contains the general introduction. Thereafter, we elaborate on social groups and, to a lesser degree, social networks. Both are the prevalent concepts to describe social fabric or structure. We then proceed to households, the social groups at the center of our analysis. We develop the formal model of households as economic decision units.

Part II: The Basic Framework. It consists of prototype general equilibrium models with multi-member households. The key concepts are introduced and discussed, and the major findings are presented. The chapter on cores provides a synopsis of several results that are related to various other chapters.

Part III: Other Forms of Group Formation and Decisions. Models of group formation are not new in microeconomic and game theory. We first relate our approach formally to alternative modeling approaches, in particular in club theory and the theory of two-sided matching. We then elaborate further on the related literature.

Part IV: Extensions and Applications. The basic framework can be extended or specialized depending on the issues and applications at hand. In those final chapters, we work out some of the applications and indicate others.

Over the years, we have received comments and encouragement from many colleagues and friends, in particular the late Birgit Grodal and in alphabetical order, Elias Aptus, Sylvain Beal, Clive Bell, Helmut Bester, Volker Böhm, Pierre-André Chiappori, Egbert Dierker, Jürgen Eichberger, Bryan Ellickson, Theresa Fahrenberger, Louis Gevers, Rob Gilles, Edward Glaeser, Ani Guerdjikova, Volker Hahn, Martin Hellwig, Stephan Imhof, Roger Lagunoff, Jean-François Mertens, Benny Moldovanu, Anne van den Nouweland, Oriana Ponta, Till Requate, Jesus Santos, Martin Scheffel, Christoph Schmidt, Klaus Schmidt, Urs Schweizer, Susan Snyder, Eva Terberger, and Bill Zame. We are thankful to all of them. We thank especially Martina Bihn from Springer-Verlag for her persistent interest in the completion of the book. Jürg Müller helped with the bibliography. Margrit Buser and Claire Burrin assisted in proofreading. Work on the book began while Hans Haller was guest professor at ETH Zürich. He is grateful to ETH for its hospitality and support and to Virginia Tech for granting a study-research leave.

Zürich, Switzerland — Hans Gersbach
Blacksburg, USA — Hans Haller

Contents

Part I
The Theme

Chapter 1
Introduction

This book introduces the reader to past and current research at the interface of general equilibrium theory and economics of the household and groups in general. It summarizes, consolidates, expands and organizes our work on general equilibrium models with multi-member households (and multi-member groups). The focus lies on the interaction between households (household formation, household decisions) and competitive markets for commodities. For that purpose, we develop a comprehensive framework that allows the integration of three allocative mechanisms involving households: (a) Household formation (and dissolution), i.e., individual decisions are made to join or leave households. (b) Household decisions, i.e., collective decisions within households regarding the individual consumption plans of household members. (c) Competitive exchange across households. Economic theorists learn which questions can be and have been addressed within this framework. They may also discover intriguing issues which remain as yet unexplored. Theorists, applied and empirical scholars alike can get additional insights from a general equilibrium approach which cannot be gained from partial equilibrium analysis.

The vast majority of humans live in households, be it households consisting of singles, couples, nuclear families, single-parent families, extended families or households different from family units, although some persons do not belong to specific households, like prison inmates and inhabitants of psychiatric institutions. We submit that the allocation of resources among consumers and the ensuing welfare properties are affected by the partition of the population into households and by the way household decisions are made. Living together in households impacts upon the welfare of individuals for various reasons. Spouses, for example, experience conjugal pleasures and displeasures. More generally, an individual's opportunities in a multi-person household depend on the resources, preferences, and actions of other household members. The kind of living quarters most couples can afford jointly differs from the separate units they could occupy as individuals. More generally, multi-member households may benefit from economies of scope and scale. Living in the same household facilitates joint activities, sharing of household chores as well as division

H. Gersbach and H. Haller, *Groups and Markets*,
DOI 10.1007/978-3-319-60516-6_1

of labor within the household, possibly risk sharing among household members. But it can also lead to certain negative externalities caused by different preferences—at least in some dimensions. To conclude, welfare analysis cannot ignore the details of household composition and decision making. Neither can empirical economics, given that the demand of multi-member households may not conform to a representative consumer model for the household. Economic policy analysis, both empirically and theoretically, from a positive and a normative perspective, has to recognize the differences across households when it comes to labor supply, consumption, savings and fertility decisions. While it is important to examine the role of households for economic outcomes, it is equally important to explore how economic conditions affect household formation, household decisions, and household stability. Apart from social norms and psychological factors, economic factors tend to play a crucial role in marriage decisions including divorce.

Traditional economic theory and empirical research have treated households as if they were single consumers. The traditional "workhorse model" of the household, as Apps and Rees (2009) call it, has been widely used and is extremely useful in consumer theory, labor economics, public economics, and other areas of inquiry. But it has its limitations on descriptive, theoretical and empirical grounds. Samuelson (1956) was first to observe that the aggregate demand function of a multi-member household can have different properties than the demand of a standard individual consumer. The crucial property is the Slutsky equation which need not hold for multi-member households. But it was Becker (1973, 1981) who started and fostered modern family economics. See Pollak's (2003) lucid account of Gary Becker's contributions to family and household economics.

We depart from traditional economic theory and allow for households with several, typically heterogeneous, members; households that make (efficient) collective consumption decisions where different households may use different collective decision mechanisms; yet households that operate within a competitive market environment. This departure from the traditional market model permits us to investigate the interplay of dual roles of households, households as collective decision making units on the one hand and households as competitive market participants on the other hand. While we maintain the term "household" throughout, the broader interpretation as socio-economic group or simply group would be appropriate in many instances, in particular since we do not impose restrictions on household or group size, respectively.

With one notable exception, we assume collective rationality of households à la Chiappori (1988b, 1992), simultaneously pioneered by Apps and Rees (1988), "… namely, that the household always reaches Pareto-efficient agreements" (Chiappori 1988a, p. 64). In our most general model, we adopt collective rationality in its broadest sense. A household chooses a consumption plan for all its members from its efficient budget set, that is the Pareto frontier of the budget set. No budget sharing rule (in the sense of Chiappori) or generalized household welfare function (in the sense of Apps and Rees (2009)) is assumed. Different households may apply different decision criteria. The reasons why we take this approach are two-fold. First of all, there is some obvious path-dependence in our research agenda: Pierre-André Chiappori gave an invited lecture on "Efficient Intra-Household Allocations: A General Characterization and Empirical Tests" at the 1994 European Meeting of the Econometric Society in Maastricht. After his presentation, Helmut Bester and Hans Haller had a conversation during which Helmut raised the question what would happen if the collective rationality model was embedded in a general equilibrium model with multi-member households. *Prima facie*, one would have expected that the welfare properties of the resulting general equilibrium allocations might depend on the particular bargaining protocols employed by households. The important insight reported in Haller (2000) is that this is not the case: Competitive exchange among given multi-member households leads to a Pareto-optimal allocation as long as each household makes an optimal (efficient) choice subject to its budget constraint and, by doing so, exhausts its budget. Haller (2000) was drafted and first presented at CentER in Tilburg in 1995. Hans Gersbach attended that talk and became interested in the topic. Soon afterwards, our collaboration began.

Inertia in model building would explain but not justify our persistent assumption of collective rationality. The second reason why we adhere to collective rationality broadly defined is that for some of the welfare analysis further details do not matter, indeed; whether there is income pooling or not, whether there is a numerical household objective function or not, etc. Also, in order to discern the impact of outside options available to household members, it proves useful to preserve utmost flexibility in modelling household decisions.

There exists by now a fair number of models of multi-member households in the literature. Browning et al. (2006) present a taxonomy that consists of four categories. Apps and Rees (2009) suggest a different classification. They offer a subtle assessment of various modelling strategies. They stress the usefulness of considering social welfare functions for the household in the spirit of Samuelson—despite Samuelson's rather critical view of family economics. Collective rationality as we see it encompasses most of the specifications of household models found in the literature. In empirical and applied work, more restrictive assumptions are necessary. We also make restrictive assumptions when warranted, for instance, when we perform comparative statics with respect to the intra-household balance of bargaining power. In Gersbach and Haller (2005, 2006) and Chap. 13, we digress and consider inefficient household decisions.

The next chapter contains an informal review of the two most prominent components of social structure studied in economics and game theory: social networks and social groups. While they are usually analyzed separately, they do, of course, co-exist. And they may be related. There may be restrictions on group and coalition formation given by a binary relation (graph, network) as in Kirman (1983), Kirman et al. (1986), Haller (1990), Gilles et al. (1994). Conversely, certain networks may be confined to particular social groups.

In Chap. 3, we develop our model of a (single-member or multi-member) household operating in a perfectly competitive market environment. In the model, all household members have their own private consumption. They care about their own consumption and possibly the consumption of other household members. Household decisions are based on two premises. First, the members of the household are subject to a joint budget constraint. This is a necessary but not sufficient condition for income pooling, a property widely debated and examined in the empirical literature. Our second premise is collective rationality of households which means that the household chooses an element at the Pareto frontier of its budget set. Collective rationality that broadly defined leaves the exact decision criterion of the household unspecified. Different households may apply different criteria.

In Chaps. 4–7, we perform general equilibrium analysis. Our goal is the integration of the three already mentioned allocative mechanisms involving households: (a) Household formation (and dissolution), i.e., individual decisions are made to join or leave households. (b) Household decisions, i.e., collective decisions within households regarding the individual consumption plans of household members. (c) Competitive exchange across households. That goal is achieved in several stages which reflect the historical development. With multi-member households, the household structure, the partition of the population into households, becomes crucial. At the initial stage presented in Chap. 4, the household structure is fixed as in Haller (2000). Hence only (b) household decisions and (c) competitive exchange across households are considered. The model analyzed in Chap. 5 incorporates production while the household structure is again fixed. Thereafter, we return to pure exchange economies. At the intermediate stage treated in Chap. 6, variable household structures are introduced. This innovation allows novel comparative statics and a more elaborate welfare analysis. Still, the model is confined to (b) and (c). The final stage is presented in Chap. 7, where we add (a) household formation and dissolution.

With a fixed household structure, the main issues are existence of competitive equilibria and the welfare properties of competitive equilibrium allocations. In Chaps. 4 and 5, we present several equilibrium existence results. The first welfare theorem of Haller (2000) is extended to finite economies with a fixed household structure and production. In several of the existence results as well as the first welfare theorem, the budget exhaustion property is assumed: A household exhausts its budget when it chooses a bundle at the Pareto frontier of its budget set. This property is often but not always satisfied. A sufficient condition are strictly monotonic preferences in own consumption and nonnegative consumption externalities. For pure exchange economies with a fixed household structure, a second welfare theorem holds, too. It is a corollary to Proposition 6.4.

With variable household structure (as in Chap. 6) or endogenous household structure (as in Chap. 7), an allocation consists of an allocation of consumers, that is a household structure, and an allocation of commodities to consumers. Accordingly, a new and more demanding optimality criterion suggests itself: A full Pareto optimum or optimum optimorum is a feasible allocation that cannot be improved upon by means of another feasible allocation consisting of a household structure and a commodity allocation. A household structure is termed optimal if it is part of a full Pareto optimum. We find that having a social planner rearrange households so that the household structure is optimal and leaving the allocation of commodities to the market need not yield a full Pareto optimum. Whereas under budget exhaustion the resulting equilibrium allocation is optimal given the household structure, a constrained Pareto optimum, the overall allocation (household structure plus commodity allocation) can fail to be fully Pareto optimal. However, a second welfare theorem holds: A fully Pareto optimal allocation is obtained as competitive equilibrium outcome after the social planner fixes the corresponding optimal household structure and redistributes endowments in a suitable way. When the household structure is endogenous, the welfare properties of competitive equilibria depend on the stability criteria imposed on households which in turn are defined in terms of the outside options household members have. Equilibria may be Pareto ranked in case the only stability requirement is that nobody can benefit from exit, that is from leaving the household and making it on their own at the going prices. Adding the further requirement that nobody can benefit from leaving their household and joining another household (without decreasing the welfare of the members of the other household) constitutes a refinement that does not necessarily eliminate inferior equilibria. Still, very strong assumptions yield strong conclusions. Core inclusion results obtain under the most stringent stability requirement that no group of consumers can benefit from forming a new household. However, such equilibria rarely exist. Equilibria need not exist even in a two-sided matching model if there are active commodity markets, contrary to the existence of stable matchings in the classical matching framework. The fact that a stable matching and market clearing cannot be achieved simultaneously is an important insight that could not be gained in models where commodity markets are absent, inactive or obsolete.

Chapter 8 is devoted to the study of several core concepts. In the pure exchange context, an improvement by a coalition upon a given allocation relies on the allocation of the coalition's aggregate endowment among its members. Thus, the coalition is treated as a sub-economy. In our context, not only are commodities allocated to consumers, but also—in the case of a variable household structure—consumers are allocated to households. In the latter case, we require that a deviating coalition viewed as a sub-economy comes up with an allocation of its resources as well as its own household structure. Although core theory is not central to most of our investigations, we have nonetheless accumulated substantial findings across a number of publications, among those several core inclusion results and an intriguing novel example of non-existence. The chapter organizes and summarizes those findings.

Household formation is an instance of group formation. Therefore, the question arises how our model of household formation is related to extant models of matching, clubs, etc. In Chap. 9, we elaborate primarily on the relationship of our approach to club theory and theories of matching. First of all, one might argue that a household is just a special kind of club. After all, the club literature for the most part deals with an endogenous partition of the population into groups, too, and some of the literature allows for the competitive market allocation of multiple private goods as well. Therefore, the distinction between "clubs" and "households" is, perhaps, purely semantic. We find that our household model and the existent club models differ in important ways, on purely descriptive grounds on the one hand and theoretical grounds on the other hand. *Prima facie*, there are a variety of descriptive features distinguishing between the club model and the household model. First, in traditional club theory, the benefit of a club to a member is determined by its membership profile and/or the provision of local public goods (club goods) or abstract club projects. In serving this purpose, the club incurs a resource cost which it tries to recoup through the collection of admission fees. Procurement of private goods remains an individual decision. In contrast, the members of a household in our model face a common budget constraint and reach a collective decision regarding the consumption of private goods. Second, clubs are formed by individual optimizing choices, regarding consumption of private goods and club membership(s). In the household model, household membership per se is not priced but the individual in a multi-member household is subject to the household's budget constraint and collectively rational consumption choice. Consequently, the expenditure on the individual's private consumption may differ from what the individual could afford as a single person.[1] Third, in club theory, externalities in private good consumption are typically absent whereas they constitute an integral part of our household model. The theoretical comparison shows that the concept of a competitive equilibrium where no group of consumers can benefit from forming a new household and the concept of valuation equilibrium used in the club literature are by and large equivalent in the absence of consumption externalities. But the equivalence breaks down in the presence of consumption externalities. Both results are shown in Gersbach and Haller (2010) and detailed in Chap. 9. In the second section of the chapter, we address the relationship of our model of household formation and the matching literature. Our general framework encompasses deterministic matching models, often with the added feature that (b) groups (households) make collective decisions regarding the individual consumption plans of their members and (c) there is competitive exchange across households. We are going to show that the existence results carry over from the matching literature to our framework when there is only one commodity. Thereafter, we are going to resume the discussion of the before-mentioned counter-example from Gersbach and Haller (2011) where there are two commodities and active trade, and stable matching and market clearing cannot occur simultaneously. The final section of Chap. 9 comments briefly on further theories of group formation.

In Chap. 10, we revisit, reorganize and reassess some of the literature cited in other chapters. Those references are combined with and related to publications that are not mentioned elsewhere in the book. Our treatment of the literature is far from comprehensive and cannot do justice to all important contributions, in particular the host of empirical work on household decisions.

Chapter 11 covers several aspects of power in general equilibrium models with multi-member households or, more precisely, power in households in a general equilibrium setting. We start with the distinction between formal and real power. The former refers to the say an individual has in a group decision, expressed, e.g., by his/her relative bargaining power. The latter refers to the utility gain an individual can obtain in a household, compared to his or her utility when being single. We study how changes of intra-household formal power in one, a few or many house-

[1] In some instances, club admission fees can be negative so that there are transfers between club members, mimicking a joint budget constraint. Still, after receiving a budget share, each club member shops for his own private consumption bundle.

holds affect the allocation of resources, as well as welfare, at both the individual and societal level. It turns out that these effects are subtle and depend on whether and how much equilibrium prices are affected by changes of formal power. In particular, higher formal power may not benefit those persons—or groups of persons—who become more powerful in a formal sense. Subsequently, we provide two ways how formal power can be endogenized and explained by primitives of the model. With the *power of voice* and the *power of (un)friendliness*, we provide two novel ways to analyze how members of a group can reach a consensus. It turns out that such concepts not only determine the allocation of resources within households, but also impact on the household structure itself. Finally, we study the relationship between formal and real power in detail. It turns out that high, or even maximal, real power is by no means an indication of Pareto inefficiency, and that price effects or a reshuffling of the household structure may translate higher formal power into lower real power.

In Chaps. 12–14, we consider several model variants to investigate specific questions. In Chap. 12, we demonstrate how the model can be utilized to evaluate the risk-sharing capacity of markets versus the risk-sharing capacity of households. We introduce uncertainty in our general equilibrium model with multi-member groups, following the classical state-space approach of Arrow-Debreu. A host of new interesting economic issues emerge. First, risk averse agents can attempt to insure themselves through markets or through mutual insurance within a multi-member group, say a household, by pooling resources within the group. Which insurance mechanism is chosen and to which extent the mechanisms substitute or complement each other is an open question. Second, one may ask more specifically what is the role of social groups for risk sharing and risk allocation when agents face idiosyncratic or aggregate risk. Third, does a suitable combination of social group formation and contingent commodity markets yield efficient risk allocations? We present a series of examples that shed some light on these issues. While the examples prove instructive, they hint only at the potential directions of future research. Many important questions remain unresolved or deserve a more systematic investigation.

Chapter 13 is devoted to general equilibrium models where household decisions can be inefficient, a digression from collective rationality. No doubt, household decision making could be prone to inefficiencies because of severe frictions, strategic behavior or simply mistakes. Then the question is how market performance is affected by inefficient household decisions. We found in Gersbach and Haller (2005, 2006) and report in Chap. 13 that one can distinguish two types of inefficiencies (mistakes): inefficient distribution of resources within the household and inefficient net trades. In the case of inefficient distribution within the household, an allocation is never Pareto optimal. In contrast, inefficient net trades may but need not impede Pareto optimal outcomes. Pareto optimal allocations can occur, if a household's mistakes are accompanied and in a sense compensated by mistakes of other households. With endogenous household formation, the competition for partners can eliminate or reduce the inefficiency of household decisions.

Our work so far deals with the interaction of three allocative mechanisms involving households: (a) Household formation (and dissolution), i.e., individual decisions are made to join or leave households. (b) Household decisions, i.e., collective decisions within households regarding the individual consumption plans of household members. (c) Competitive exchange across households. In Chap. 14, we suggest the addition of a fourth mechanism, public choice regarding the provision and funding of public goods. We outline a general framework and suggest a number of applications, some of which we touch upon in Gersbach and Haller (2014), but most of which are unexplored.

In Chap. 15, we indicate how our models could be used for economic policy analysis. Some of our theoretical results have immediate policy implications. In other instances, parametric versions of the model might prove very useful. Apps and Rees (2009) and articles of theirs analyze public economics and taxation under the assumption that the bulk of households are one-adult or two-adult households, with or without children. This goes beyond the traditional single-person model. Still, additional insights could be gained from a general equilibrium perspective and from models with substantial heterogeneity of household types. We take the general equilibrium approach to investigate the interplay of household formation, household decisions and competitive exchange of commodities. We allow different households (possibly of similar composition) to use different decision criteria. It may well be that ample heterogeneity in the real economy explains some of the inconclusiveness of the empirical evidence.

We trust to have shown that the analysis of multi-member households in a general equilibrium setting is worth the effort. But we would hope and are confident that this endeavor does not end with this book. There are many ramifications and unexplored venues for future investigations. To name just a few: In a first pass on general equilibrium models with multi-member households, we stuck with collective rationality of households, with the exception of Chap. 13. More specific instances of inefficient household decisions ought to be scrutinized with respect to their general equilibrium implications, for example separate spheres bargaining à la Lundberg and Pollak (1993). Household production, despite its declining importance, could be incorporated. Alternative outside options might be considered. We assume that a person when leaving a household has the option to be single which is typically the case in the societies we live in. However, this is not the case in other societies. A more systematic study of two-sided matching with active commodity markets could be envisaged. A general equilibrium perspective of the taxation of multi-member households, alluded to in the previous paragraph, is of utmost importance.

Introducing durability in household formation would necessitate an intertemporal setting which could follow standard dynamic macroeconomic approaches to put time at the center of analysis of general equilibrium theory. This focus on durable households (or firms) is an important route that remains to be explored. Combined with the explicit incorporation of children and their special role and status in households, such exploration may provide a more comprehensive picture of the forces holding a social structure together.

Finally, many of the ideas presented in this book may be of use to take up issues beyond the confines of the particular models we are going to analyze. Let us illustrate that point by means of four examples. First, the interaction between different allocative mechanisms (e.g., collective decisions, group formation and competitive exchange) is important in its own right and may prove useful in other contexts such as the formation of firms, which we have barely examined so far. Second, the way collective decisions on public goods are taken and how competitive markets with multiple commodities operate is of fundamental importance for the functioning of modern societies, which are typically governed by those two allocative (and distributive) mechanisms. The approach outlined in Chap. 14 could be taken much farther. Third, the evolution of societies often exhibits a great degree of path-dependency—with current physical, institutional and belief conditions determining the scope of future development. Household and firm structures, the backbones of society, may thus display path-dependency. Dynamic features of this nature could be added to our framework. Fourth, we have introduced new concepts such as the *power of voice* or the *power of (un)friendliness*. They constitute widely applicable approaches to endogenize bargaining power. The basic ideas can be applied to any circumstances in which bargaining takes place. Moreover, these concepts are merely the catalyst for a research program that promises considerable novel insights into interaction between individuals. Furthermore, the application of the *power of voice* to politics could open up new ways to formalize deliberation in democracy. It may provide the foundation for well-functioning democracies or at least better working democracies—beyond basic principles such as equal voting and agenda-setting rights, elections, separation of powers, independent judicial systems, the protection of liberty, and human rights. To conclude, the examples suggest that a number of our ideas transcend the framework to be delineated in the remaining chapters.

References

Apps, P. F., & Rees, R. (1988). Taxation and the household. *Journal of Public Economics, 35*, 355–369.

Apps, P. F., & Rees, R. (2009). *Public economics and the household*. Cambridge, UK: Cambridge University Press.

Becker, G. S. (1973). A theory of marriage, Part I. *Journal of Political Economy*, *81*, 813–846. Reproduced as Chap. 11, in R. Febrero & P. S. Schwartz (Eds.), *The Essence of Becker*. Stanford, CA: Hoover Institution Press, 1995.

Becker, G. S. (1981). *A treatise on the family*. Cambridge, MA: Harvard University Press.

Browning, M., Chiappori, P., & Lechene, V. (2006). Collective and unitary models: A clarification. *Review of Economics of the Household*, *4*, 5–14.

Chiappori, P.-A. (1988a). Nash-Bargained household decisions: A comment. *International Economic Review*, *29*, 791–796.

Chiappori, P.-A. (1988b). Rational household labor supply. *Econometrica*, *56*, 63–89.

Chiappori, P.-A. (1992). Collective labor supply and welfare. *Journal of Political Economy*, *100*, 437–467.

Chiappori, P.-A., & Ekeland, I. (2006). The micro economics of group behavior: General characterization *Journal of Economic Theory*, *130*, 1–26.

Gersbach, H., & Haller, H. (2005). When inefficiency begets efficiency. *Economic Theory*, *25*, 105–121.

Gersbach, H., & Haller, H. (2006). Household inefficiency and equilibrium efficiency. In C. Schultz & K. Vind (Eds.), *Institutions, equilibria and efficiency: Essays in honor of Birgit Grodal* (pp. 187–209). Heidelberg: Springer-Verlag.

Gersbach, H., & Haller, H. (2010). Club theory and household formation. *Journal of Mathematical Economics*, *46*, 715–724.

Gersbach, H., & Haller, H. (2011). Competitive markets, collective decisions and group formation. *Journal of Economic Theory*, *146*, 275–299.

Gersbach, H. & Haller, H. (2014). Households, markets and public choice. CESifo Working Paper 4947.

Gilles, R. P., Haller, H., & Ruys, P. H. M. (1994). Modelling of economies with relational constraints on coalition formation. In R. P. Gilles & P. H. M. Ruys (Eds.), *Imperfections and behavior in economic organizations, Chap. 5*. Boston, MA: Kluwer Academic Publishers.

Haller, H. (1990). Large random graphs in presudo-metric spaces. *Mathematical Social Sciences*, *20*, 147–164.

Haller, H. (2000). Household decisions and equilibrium efficiency. *International Economic Review*, *41*, 835–847.

Kirman, A. (1983). Communication in markets: A suggested approach. *Economics Letters, 12*, 101–108.

Kirman, A., Oddou, C., & Weber, S. (1986). Stochastic communication and coalition formation. *Econometrica, 54*, 129–138.

Lundberg, S., & Pollak, R. A. (1993). Separate spheres bargaining and the marriage market. *Journal of Political Economy*, *101*, 988–1011.

Pollak, R. A. (2003). Gary Becker's contributions to family and household economics. *Review of Economics of the Household*, *1*, 111–141.

Samuelson, P. A. (1956). Social indifference curves. *Quarterly Journal of Economics*, *70*, 1–22.

Chapter 2
Social Groups

Social fabric and structure are by and large described by means of social relations and networks, like kinship, and by group affiliation such as household membership or club membership. To accentuate special social features or to investigate specific societal issues, one descriptive tool, relations and networks, or the other, group membership, or a combination of both will be appropriate.

Social Networks. The main subjects of our inquiry are going to be household decisions and household formation in a competitive market environment or, more generally, collective decisions by groups and group formation under competitive market conditions. Therefore, our treatment of networks will be rather short despite the rapidly growing importance of networks and network analysis—which is now recognized in many fields, for instance in artificial intelligence, biology, business and finance, computer science, economics, electrical engineering, neuroscience, sociology, and physics.

Network analysis can be focused on network topology, network utilization, network formation, or the co-evolution of network utilization and network formation. Within game theory, several strands of literature on network creation (network formation, network design) have emerged. A number of recent contributions have treated social and economic networks as the outcome of a network formation game. The players of the game constitute the nodes of the network to be formed. In the purely non-cooperative approach of Bala and Goyal (2000), addition and deletion of links are unilateral decisions of the player from whom the respective links originate. The player's strategy is a specification of the set of agents with whom he forms links. The costs of link formation are incurred only by the player who initiates the link. The formed links define the network. Pairwise stability à la Jackson and Wolinsky (1996) treats addition of a link in a network as a bilateral decision by the two players involved, whereas severance of a link constitutes a unilateral decision. The costs of a link are borne by both players constituting the link. Consensual link formation can be and has also been formulated and studied in a purely non-cooperative context,

H. Gersbach and H. Haller, *Groups and Markets*,
DOI 10.1007/978-3-319-60516-6_2

like in the seminal contribution by Myerson (1991, p. 448). For surveys on strategic network formation, see Goyal (2007), and Jackson (2005, 2008).

Local interaction or spatial games capture important aspects of network utilization. In a static local interaction game, players exhibit myopia in the spatial sense which means that the player's payoff is affected only by the strategic choices of his immediate neighbors. Most models in that literature, e.g. Berninghaus and Schwalbe (1996) and Blume (1995) also incorporate a dynamic or evolutionary element. In best response dynamics, at each time, one or every player plays a (static) best response against the empirical distribution of the last strategies chosen by his neighbors. This constitutes rational behavior impaired by spatial and temporal myopia. Myopia in the temporal sense means that the player is not forward looking, and does not take into account that other players might be changing their strategies as well. Furthermore, the literature considers small random deviations from best response play and corresponding notions of stochastic stability, for instance Baron et al. (2002), Berninghaus et al. (2006), Ellison (1993), and Young (1998, Chap. 6). The main alternative to best response dynamics are various types of imitation dynamics, for example in Eshel et al. (1998), Josephson and Matros (2004), Kirchkamp (2000), Nowak and May (1993) and Outkin (2003).

While the focus on either network formation or network utilization provides valuable insights, network design and network utilization may go hand in hand. The co-evolution of networks and their use has been modeled in Jackson and Watts (2002), Goyal and Vega-Redondo (2005), Hojman and Szeidl (2006), and Ehrhard et al. (2008), among others. Finally, while our focus will be solely on social groups, social groups and social networks frequently co-exist and may influence, presuppose or even cause each other. For instance, a member of a social network (modeled as a graph) plus that person's immediate neighbors form a specific social group, say the person's reference group in a model of opinion formation. Conversely, common membership in some socio-economic group may be a prerequisite or catalyst for two individuals to be linked in a particular social network. For example, with rigid social stratification, kinship may only be possible or likely within the same social class, caste, etc.

Social Groups. Formation, composition and behavior of groups are central topics of anthropology, sociology, economics, political science, and other social sciences. Members of a society or population can be classified and categorized in many ways, for instance by age, height, weight, race, gender, marital status, education, occupation, income, place of birth, place of residence, citizenship, language, or religion. Classification of individuals according to such attributes puts each individual into at least one class, category or group and sometimes into several categories, for example in the case of dual citizenship. While some attributes like age are innate, others are acquired and at least partially the consequence of own choices and decisions of others. While some common attributes like common language or common location facilitate social and economic interaction, they need not lead to interaction. This may be the case even if a person voluntarily joins a special interest group like the American Automobile Association (AAA). A member is entitled but not obliged to use certain services provided by the AAA, may not interact with any other AAA member, and simply be satisfied to support a "good cause", say particular lobbying activities.

Sociologists have forwarded a number of definitions of a "social group", most of which imply drastic restrictions on group size. A widespread but by no means ubiquitous definition is the following, with minor variations: *"To sociologists a group is a collection of individuals who • interact and communicate with each other; • share goals and norms; and • have a subjective awareness of themselves as "we," that is, as a distinct social unit."*[1] In contrast, a social category is a collection of people who share similar attributes or characteristics but may not all interact with each other, for example men, women, the elderly, social security recipients, the urban youth, generation X, or generation Y. Almost every social category qualifies as a social group in the terminology of many social psychologists who tend to use a very broad definition of "social groups".[2]

In economics, to the extent that the term is used at all, "social group" often refers to social categories at large. For instance, the economic analysis of the returns to education may deal with large jurisdictions, school districts, student and parent populations, teachers, and other stake-holders like tax paying local property owners and educational scholars. The term "socio-economic group" indicates or emphasizes economic activities and interactions of a group, for instance a trade union, rather than merely social ones. While some of these groups or categories can comprise thousands or millions of members, most socio-economic groups, mainly families and households, are very small.

Firms. Some cooperatives and most productive partnerships such as law firms clearly constitute socio-economic groups. In general, the nature and comprehensive description of a firm can be extremely complex. The socio-economic group aspect of the firm is but one of several characteristic features.

- A technological description deals with the firm as a production facility, as a set of feasible input-output combinations.
- Contract theory views the firm as a nexus of contractual arrangements such as employment contracts, procurement contracts, delivery contracts, financial contracts. According to some authors, that is all there is to it.
- From a Coasian or organizational perspective, the firm is a nexus of relations, exemplified by an information and communication structure, production lines and lines of command, decision-making units and processes.

Households. The household is commonly considered the basic unit of economic activity. Formal definitions of households differ across countries and disciplines, for instance the definition of households for tax purposes and the definition of households for census and demographic purposes. Some socio-economic entities clearly qualify as households: Nuclear families and married couples living together, single persons living alone, single parent homes. From an economic perspective, cohabiting partners (plus, if applicable, common children sharing their home) qualify as well. In other

[1] Andersen and Taylor (2011), p. 109.

[2] See Table 1.1 in Stangor (2004).

cases like blended families, the household affiliation of some of the members may be more difficult to determine. If non-married or separated parents have joint custody and their child spends equal time in both homes, then arguably the child should be considered part of both households. If members of Congress share an apartment in Washington, D.C., they probably spend more time and meals together than with their families back home. Still, one tends to consider the respective family as the congress persons's household rather than the living arrangement in D.C. Despite these potential ambiguities, we shall always assume that each consumer belongs to exactly one household. That is, there exists a partition of the population into households. We call such a partition a household structure.

Households in General Equilibrium. Partial equilibrium analysis has produced countless theoretical and empirical studies of household related issues, involving numerous economic sub-disciplines and touching upon topics as diverse as fertility, mortality, demography, population dynamics, marriage and matching, status, income, poverty, nutrition, health, public transfers, education, social capital, human capital, employment, development, welfare, demand and supply, and so forth. Each of the sub-disciplines has developed its own rich body of theories and accumulated a host of empirical work.

Our approach differs from partial equilibrium analysis. We take a general equilibrium perspective that allows the synopsis of three interacting allocation mechanisms, each operating at a particular level of aggregation: Individual decisions are made to join or leave households. Collective decisions within households determine the consumption plans of household members. Competitive exchange across households achieves a feasible allocation of resources. Our investigation of general equilibrium models with multi-member households is motivated by the fact that the allocation of resources among consumers and the ensuing welfare properties are obviously affected by the specifics of a pre-existing partition of the population into households (household structure) and that conversely, the formation of households can—partly or fully—be driven by economic considerations, by the anticipated effects of the emerging household structure on the allocation of economic resources.

The traditional general equilibrium model of a pure exchange economy—and traditional economic theory at large—has treated households as if they were single consumers. When we consider households in the sequel, we depart from traditional economic theory and allow for households with several, typically heterogeneous, members; households that make (efficient) collective consumption decisions where different households may use different collective decision mechanisms; yet households that operate within a competitive market environment. The distinction between a household and its members potentially leads to inquiries into household decisions, household formation, household stability, the interaction between the competitive market allocation of private goods and household formation—and to a host of related modeling issues. The departure from the traditional market model with "unitary households" permits us to investigate the interplay of the dual role of households, households as collective decision making units on the one hand and households as competitive market participants on the other hand. While we maintain the term "household" throughout, the broader interpretation as socio-economic group or simply group would be appropriate in many instances, in particular since we do not impose restrictions on household or group size, respectively.

Household Decisions. Household decisions with an economic impact are manifold. The *Review of Economics of the Household* delineates the scope of household decisions as follows: "Household decisions analyzed in the journal include consumption, labor supply and other uses of time, household formation and dissolution, demand for health and other forms of human capital, fertility and investment in children's human capital, demand for environmental and other public goods, migration, demand for religiosity, and decisions by agricultural households." This list is far from being exhaustive.

Household decisions have been widely studied in the empirically oriented literature. Of particular interest for our purposes is the contribution of Chiappori (1988, 1992) who introduced a model of collective rationality (efficient consumption decisions) of multi-member households. Haller (2000) pioneered the study of general equilibrium implications of competitive exchange among multi-member households whose decision making is described by the collective rationality model in its most general form. He assumes a given household structure, that is a partition of the population into households, and addresses the optimality of competitive exchange under these circumstances. He finds that equilibrium outcomes are Pareto optimal as long as each household makes an optimal (efficient) choice subject to its budget constraint and, by doing so, exhausts its budget. Further details of efficient decision making within households like specifics of the bargaining protocol prove irrelevant for the conclusion. Haller identifies consumption externalities within households for which the budget exhaustion property obtains.

Household Formation. Formal models of household or group formation have existed prior to our work. What is new is the integrated view of three allocative mechanisms involving households: (a) Household formation (and dissolution), i.e., individual decisions are made to join or leave households. (b) Household decisions, i.e., collective decisions within households regarding the individual consumption plans of household members. (c) Competitive exchange across households. We aim to incorporate all three facets of households into the inquiry of which households form and, consequently, which household structure prevails.

Gary Becker (1978, 1981) constitutes the most prominent early voice on endogenous household formation. He and Pierre-André Chiappori have been the greatest inspiration for our research. But we use a different model and address questions different from Becker's. For instance, household-specific externalities play an important role in our approach. In contrast, Becker's model avoids consumption externalities in a unique way, by introducing a "household good", the sole explicit consumption good which is non-tradable, yet perfectly divisible within each household and does not cause any consumption externalities.

Household formation or, more generally, group formation is the main subject of the literature on matching, assignment games, and hedonic coalitions. Our general framework includes models of two-sided matching as a special case. But it incorporates further aspects of group formation and group decisions, in particular (c) comptetive exchange between groups (households) and the integrated treatment of the three allocative mechanisms (a), (b) and (c) mentioned before. In Gersbach and Haller (2011), we present an example with two private goods and household formation reducible to a two-sided matching problem, that puts the traditional literature on matching into perspective: Stable matchings and market clearing cannot be achieved simultaneously. This kind of non-existence is notably absent from the vast majority of the matching literature, where markets are inactive and relative prices are irrelevant, simply because there exists at most one tradable commodity. Most of the work on hedonic coalitions (e.g. Banerjee et al. (2001), Bogomolnaia and Jackson (2002)), matching (e.g. Alkan (1988), Gale and Shapley (1962), Roth and Sotomayor (1990)), assignment games (e.g. Roth and Sotomayor (1990), Shapley and Shubik (1972)), and multilateral bargaining (e.g. Bennett (1988, 1997), Crawford and Rochford (1986), Rochford (1984)) focuses on group formation and lacks competitive markets for commodities. Consequently, this literature fails to observe that in general, stable matchings and market clearing cannot be achieved simultaneously. A noteworthy exception are Drèze and Greenberg (1980) who combine the concepts of individual stability and price equilibrium, but confine the analysis of their most comprehensive model to an instructive example.

Most of the club literature presumes a partition of the population into groups (clubs) and endogenous group formation as well. The more recent contributions allow also for multiple private commodities. However, there are no externalities in private good consumption and the procurement of private goods remains an individual decision. See for instance Gilles and Scotchmer (1997). The main purpose of clubs is the provision of club goods (local public goods) or club projects whose costs are recouped through the collection of admission fees. Consumers also care about the membership profile of their club. In contrast, our model of the household permits consumption externalities within households and assumes collective decisions regarding the individual consumption bundles of household members. Although all commodities are treated as private goods, we can accommodate local public goods via intra-household externalities by having individual welfare solely depend on the aggregate consumption of the good within the household. Household membership *per se* is not priced but the individual in a multi-member household is subject to the household's budget constraint and collectively rational consumption choice. Consequently, the expenditure on the individual's private consumption may differ from what the individual could afford as a single person. In Gersbach and Haller (2010), we clarify in a more systematic way the relationship between the general equilibrium model with multi-member households and club models with multiple private goods. A summary is provided in Chap. 9 below.

Integrating Three Allocation Mechanisms. Our general framework allows the integration of the three allocation mechanisms (a)–(c) operating at different levels of aggregation. It has been developed in several steps. In Haller (2000) and Gersbach and Haller (2001) we take a first step and incorporate the collective rationality concept of Chiappori (1988, 1992) into a general equilibrium framework. This setting has allowed us to study the interaction between two of the three allocation mechanisms: (b) collective decisions and (c) competitive markets. Haller (2000) assumes an exogenously given household structure. Every household member consumes an individual bundle of private goods and has individual preferences. Preferences permit positive or negative intra-household externalities: Individual welfare can be affected by own consumption and the consumption of fellow household members. Gersbach and Haller (2001) introduce a variable household structure, with household specific preferences: An individual cares about who belongs to his household and who consumes what in his household. Hence, in general, there can be group externalities (related to household composition) as well as consumption externalities (related to household consumption). An allocation consists of two parts, an allocation of commodities to consumers and an allocation of people into households. In Gersbach and Haller (2010, 2011) we take further steps towards (c) an endogenous household structure, by amending the equilibrium conditions with stability requirements known from the matching literature.

Adding a Fourth Allocation Mechanisms. In our most recent work, we have added a fourth allocation mechanism: global collective decisions regarding public good provision and redistribution—thus defining the activities of the state in a jurisdiction. In this fourth allocation mechanism, household members act as citizens and decide (individually) how to cast their vote in referenda or in elections of candidates for public office.

Such an integration of household formation, local collective decisions, competitive markets[3] and global collective decisions is arguably the most general and comprehensive challenge for any economic system, as put forward in the literature. Chapter 14 outlines the scope and potential of such a comprehensive model.

References

Alkan, A. (1988). Nonexistence of stable threesome matchings: Note. *Mathematical Social Sciences, 16*, 201–209.
Andersen, M. L., & Taylor, H. F. (2011). *Sociology: The essentials* (6th ed.). Belmont, CA: Wadsworth.
Bala, V., & Goyal, S. (2000). A non-cooperative model of network formation. *Econometrica, 68*, 1181–1229.
Banerjee, S., Konishi, H., & Sönmez, T. (2001). Core in a simple coalition formation game. *Social Choice and Welfare, 18*, 135–153.
Baron, R., Durieu, J., Haller, H., & Solal, P. (2002). Control costs and potential functions for spatial games. *International Journal of Game Theory, 31*, 541–561.
Becker, G. S. (1978). *The economic approach to human behavior*. Chicago, IL: University of Chicago Press.
Becker, G. S. (1981). *A treatise on the family*. Cambridge, MA: Harvard University Press.
Bennett, E. (1988). Consistent bargaining conjectures in marriage and matching. *Journal of Economic Theory, 45*, 392–407.
Bennett, E. (1997). Multilateral bargaining problem. *Games and Economic Behavior, 19*, 151–179.
Berninghaus, S. K., & Schwalbe, U. (1996). Conventions, local interaction, and automata networks. *Journal of Evolutionary Economics, 6*, 297–312.

[3]In some variants we also include production.

Berninghaus, S., Haller, H., & Outkin, A. (2006). Neural networks and contagion. *Revue d'Économie Industrielle, 114*(115), 205–224.

Blume, L. E. (1995). The statistical mechanics of best-response strategy revisions. *Games and Economic Behavior, 11*, 111–145.

Bogomolnaia, A., & Jackson, M. O. (2002). The stability of hedonic coalition structures. *Games and Economic Behavior, 38*, 201–230.

Chiappori, P.-A. (1988). Rational household labor supply. *Econometrica, 56*, 63–89.

Chiappori, P.-A. (1992). Collective labor supply and welfare. *Journal of Political Economy, 100*, 437–467.

Crawford, V. P., & Rochford, S. C. (1986). Bargaining and competition in matching markets. *International Economic Review, 27*, 329–348.

Drèze, J., & Greenberg, J. (1980). Hedonic coalitions: Optimality and stability. *Econometrica, 48*, 987–1003.

Ehrhard, G., Marsili, M., & Vega-Redondo, F. (2008). Emergence and resilience of social networks: A general theoretical framework. *Annales d'Économie et de Statistique, 86*, 1–13.

Ellison, G. (1993). Learning, local interaction, and coordination. *Econometrica, 61*, 1047–1071.

Eshel, I., Samuelson, L., & Shaked, A. (1998). Altruists, egoists, and hooligans in a local interaction model. *American Economic Review, 88*, 157–179.

Gale, D., & Shapley, L. (1962). College admissions and the stability of marriage. *American Mathematical Monthly, 92*, 261–268.

Gersbach, H., & Haller, H. (2001). Collective decisions and competitive markets. *Review of Economic Studies, 68*, 347–368.

Gersbach, H., & Haller, H. (2010). Club theory and household formation. *Journal of Mathematical Economics, 46*, 715–724.

Gersbach, H., & Haller, H. (2011). Competitive markets, collective decisions and group formation. *Journal of Economic Theory, 146*, 275–299.

Gilles, R. P., & Scotchmer, S. (1997). Decentralization in replicated club economies with multiple private goods. *Journal of Economic Theory, 72*, 363–387.

Goyal, S. (2007). *Connections: An introduction to the economics of networks*. Princeton, NJ: Princeton University Press.

Goyal, S., & Vega-Redondo, F. (2005). Network formation and social coordination. *Games and Economic Behavior, 50*, 178–207.

Haller, H. (2000). Household decisions and equilibrium efficiency. *International Economic Review, 41*, 835–847.

Hojman, D. A., & Szeidl, A. (2006). Endogenous networks, social games, and evolution. *Games and Economic Behavior, 55*, 112–130.

Jackson, M. O. (2005). A survey of models of network formation: Stability and efficiency. In G. Demange & M. Wooders (Eds.), *Group formation in economics; networks, clubs and coalitions, Chap. 1* (pp. 11–57). Cambridge, UK: Cambridge University Press.

Jackson, M. O. (2008). *Social and economic networks*. Princeton, NJ: Princeton University Press.

Jackson, M. O., & Watts, A. (2002). On the formation of interaction networks in social coordination games. *Games and Economic Behavior, 41*, 265–291.

Jackson, M. O., & Wolinsky, A. (1996). A strategic model of economic and social networks. *Journal of Economic Theory, 71*, 44–74.

Josephson, J., & Matros, A. (2004). Stochastic imitation in finite games. *Games and Economic Behavior, 49*, 244–259.

Kirchkamp, O. (2000). Spatial evolution of automata in the prisoners' dilemma. *Journal of Economic Behavior and Organization, 43*, 239–262.

Myerson, R. B. (1991). *Game theory: Analysis of conflict*. Cambridge, MA: Harvard University Press.

Nowak, M. A., & May, R. M. (1993). The spatial dilemmas of evolution. *International Journal of Bifurcation and Chaos, 3*, 35–78.

Outkin, A. V. (2003). Cooperation and local interactions in the prisoners' dilemma game. *Journal of Economic Behavior and Organization*, *52*, 481–503.
Rochford, S. C. (1984). Symmetrically pairwise-bargained allocations in an assignment market. *Journal of Economic Theory*, *34*, 262–281.
Roth, A. E., & Sotomayor, M. A. O. (1990). *Two-sided matching: A study in game-theoretic modeling and analysis*. Cambridge, UK: Cambridge University Press.
Shapley, L. S., & Shubik, M. (1972). The assignment game I: The core. *International Journal of Game Theory*, *1*, 111–130.
Stangor, C. (2004). *Social groups in action and interaction*. New York, NY: Psychology Press.
Young, H. P. (1998). *Individual strategy and social structure: An evolutionary theory of institutions*. Princeton, NJ: Princeton University Press.

Chapter 3
Households

Official definitions of households differ across countries and depending on the purpose of classification, sometimes even within countries. The United States 2010 Census defines households as follows: "*A household includes all the people who occupy a housing unit. (People not living in households are classified as living in group quarters.) A housing unit is a house, an apartment, a mobile home, a group of rooms, or a single room that is occupied (or if vacant, is intended for occupancy) as separate living quarters. Separate living quarters are those in which the occupants live separately from any other people in the building and which have direct access from the outside of the building or through a common hall. The occupants may be a single family, one person living alone, two or more families living together, or any other group of related or unrelated people who share living arrangements.*"
(2010 Census Summary File 1, prepared by the U.S. Census Bureau, 2011, http://www.census.gov/prod/cen2010/doc/sf1.pdf#page=504).

Households may but need not contain families. The formal definitions of families, let alone informal definitions, differ widely. We shall adhere to the term "household" and assume that each person belongs to exactly one household. Thus we ignore persons not living in households, for instance prison inmates. When we consider endogenous household formation, individuals will only form a household and stay in the household if it is to their advantage or at least, they are not disadvantaged by household membership. Advantageous household membership for everybody belonging to the household necessitates some positive externality attributable to household membership. Following the taxonomy of Gori and Villanacci (2011), we can distinguish between three types of externalities:

- Consumption externalities where an individual's welfare is affected by the consumption of other household members.
- Group externalities where an individual's welfare is affected by the identity of other household members.

H. Gersbach and H. Haller, *Groups and Markets*,
DOI 10.1007/978-3-319-60516-6_3

- Endowment externalities where the household endowment differs from the sum of the endowments that each member would have as a single.

If several externalities are present, they can interact and may jointly reinforce or weaken the benefits of being a member in a household. Most of our analysis deals only with consumption and group externalities. In some of our work, we also cover endowment externalities. Those externalities allow to apply the model to productive partnerships and relation-specific investments by its members.[1] In the description of the model we use the term "households" that can, however, encompass such partnerships.

Consumers and Households. We consider a finite population of **consumers or individuals**, represented by a set $I = \{1, \ldots, n\}$ with $n > 1$. A generic consumer is denoted i or j. A group of consumers or (potential) **household** is any non-empty subset h of the population I. A generic household is denoted h or g. $\mathcal{H} = \{h \subseteq I | h \neq \emptyset\}$ denotes the set of all potential households. For $i \in I$, $\mathcal{H}_i = \{h \subseteq I | i \in h\}$ denotes the set of all potential households which have i as a member. While we stick to the suggestive term "household", a broader interpretation as socio-economic group or simply group would be quite appropriate in many instances, in particular, since as a rule, we do not impose any restrictions on household or group size, respectively. The latter does not mean that arbitrary households will form. A potential household will not exist in equilibrium if some of its member have better alternatives. For instance, membership in extremely large households may be unattractive compared to living in smaller ones.

The population I is partitioned into households: There exists a partition P of I into non-empty subsets referred to as households. For a consumer $i \in I$, $P(i)$ denotes the unique element of P (unique household in P) to which i belongs. If P consists of H households, we frequently label them $h = 1, \ldots, H$, provided this causes no confusion.

We call any partition P of I a **household structure in** I. We treat the household structure as an object of endogenous choice. Households are endogenously formed so that some household structure P is ultimately realized. Consequently, our **consumer allocation space** is $\mathcal{P}$, the set of all household structures in I.

Relative to a household structure P, we use the following terminology regarding $i \in I$ and $h \subseteq I$, $h \neq \emptyset$:

"household h exists" or "household h is formed"	iff $h \in P$;
"i belongs to h" or "individual i is a member of household h"	iff $i \in h$.

Commodities. With the exception of Chap. 14, each commodity is formally treated as a private good, possibly with externalities in consumption. There exists a finite number $\ell \geq 1$ of such commodities. Thus the commodity space is $\mathbb{R}^\ell$. Consumer $i \in I$ has consumption set $X_i = \mathbb{R}^\ell_+$ so that the **commodity allocation space** is

[1] This theme has been developed in Gersbach and Haller (2004).

$\mathcal{X} \equiv \prod_{j \in I} X_j$. Generic elements of $\mathcal{X}$ are denoted $\mathbf{x} = (x_i)$, $\mathbf{y} = (y_i)$. Commodities are denoted by superscripts $k = 1, \ldots, \ell$. For a potential household $h \subseteq I$, $h \neq \emptyset$, set $\mathcal{X}_h = \prod_{i \in h} X_i$, the consumption set for household h. $\mathcal{X}_h$ has generic elements $\mathbf{x_h} = (x_i)_{i \in h}$. If $\mathbf{x} = (x_i)_{i \in I} \in \mathcal{X}$ is a commodity allocation, then consumption for household h is the restriction of $\mathbf{x} = (x_i)_{i \in I}$ to h, $\mathbf{x_h} = (x_i)_{i \in h}$.

Endowments. The economic units endowed with resources are households rather than individuals. Note, however, that in an environment with endogenous household formation, each singleton $\{i\}$ is a potential one-person household with its own endowment.

For a potential household $h \subseteq I$, $h \neq \emptyset$, its **endowment** is a commodity bundle $\omega_h \in \mathbb{R}^\ell$, $\omega_h \geq 0$. In general, the social endowment with resources depends on the household structure. Namely, if the household structure $P \in \mathcal{P}$ is in place, then the **social endowment** is

$$\omega_P \equiv \sum_{h \in P} \omega_h.$$

A different household structure can yield a different social endowment. Allowing the endowment of a household to differ from the sum of endowments of the potential one-person households formed by its members can be interpreted as resource costs of setting up households or, in the opposite direction, as economies of scale enjoyed by larger households.

A special case is

(IPR) Individual Property Rights: $\omega_h = \sum_{i \in h} \omega_{\{i\}}$ for each household h.

(IPR) amounts to absence of endowment externalities in all potential households and is frequently though not always assumed in our analysis. Under (IPR), the social endowment is independent of the household structure and equals $\omega_S = \sum_{i \in I} \omega_{\{i\}}$. In fact, (IPR) holds if and only if the social endowment is independent of the household structure.

Allocations. An **allocation** is a pair $(\mathbf{x}; P) \in \mathcal{X} \times \mathcal{P}$ specifying the consumption bundle and household membership of each consumer. We call an allocation $(\mathbf{x}; P) \in \mathcal{X} \times \mathcal{P}$ **feasible**, if

$$\sum_{i \in I} x_i = \omega_P. \tag{3.1}$$

After the specification of individual preferences, by means of utility representations, an allocation determines the welfare of each and every member of society.

Consumer Preferences. In principle, a consumer might have preferences on the allocation space $\mathcal{X} \times \mathcal{P}$ and care about each and every detail of an allocation. But we shall restrict our analysis to situations of **household-specific preferences** where the consumer does not care about the features of an allocation beyond the boundaries of his own household. If a particular household structure is given, he is indifferent about the affiliation and consumption of individuals not belonging to his own household. That is, consumer $i \in I$ is indifferent between two allocations $(\mathbf{x}; P)$ and $(\mathbf{x}'; P')$ if there exists a household h such that $h = P(i) = P'(i)$ and $\mathbf{x_h} = \mathbf{x}'_\mathbf{h}$. We are going to make the **Assumption of Household-Specific Preferences (HSP)** throughout the book. It proves convenient to represent such preferences by utility functions. To this end, let us denote $\mathcal{X}^* = \bigcup_{h\in\mathcal{H}} \mathcal{X}_h$ and define $\mathcal{A}_i = \{(\mathbf{x_h}; h) \in \mathcal{X}^* \times \mathcal{H} : h \in \mathcal{H}_i,\ \mathbf{x_h} \in \mathcal{X}_h\}$ for $i \in I$. We assume that each individual $i \in I$ has a UTILITY REPRESENTATION $U_i : \mathcal{A}_i \to \mathbb{R}$.

The assumption (HSP) is justifiable on the grounds that we want to design a model where multi-member households play a significant allocative role. With a fixed, exogenously given household structure P, household membership can be considered part of an individual's identity in which case household affiliation may be dropped as argument of the individual's utility function. Thus, one obtains the utility representation $u_i : \mathcal{X}_h \to \mathbb{R}$, given by $u_i(\mathbf{x_h}) = U_i(\mathbf{x_h}; h)$ for $\mathbf{x_h} \in \mathcal{X}_h$ and $i \in h \in P$.

(HSP) still admits a lot of flexibility. For example, it permits various kinds of consumption externalities within households. In particular, (HSP) allows to accommodate the presence of local public goods within a household, although all commodities are treated as private goods. Good k is de facto a local public good for household h if the individual welfare of each household member solely depends on the aggregate consumption of the good within the household: There exist functions $V_i : \mathbb{R}_+^{|h|(\ell-1)+1} \times \{h\} \to \mathbb{R},\ i \in h$, such that

$$U_i(\mathbf{x_h}; h) = V_i((x_j^l)_{j\in h,\, l\neq k}, \textstyle\sum_{j\in h} x_j^k; h) \text{ for } \mathbf{x_h} \in \mathcal{X}_h, i \in h.$$

Suitable externalities may prevent (or foster) the formation of certain households, even though we are not explicitly restricting household size and household profiles. In the sequel, we shall in particular exploit the occurrence of pure group externalities that depend solely on the persons belonging to a household, not on what they consume. Pure group externalities capture in reduced form various aspects of the goods and bads of human beings living together.

(PGE) Pure Group Externalities: For each consumer i, there exist functions $U_i^c : X_i \to \mathbb{R}$ and $U_i^g : \mathcal{H}_i \to \mathbb{R}$ such that $U_i(\mathbf{x_h}; h) = U_i^c(x_i) + U_i^g(h)$ for $\mathbf{x_h} \in \mathcal{X}_h, h \in \mathcal{H}_i$.

(PGE) assumes that one can additively separate the pure consumption effect $U_i^c(x_i)$ from the pure group effect $U_i^g(h)$. A very special case is the **absence of externalities**, corresponding to $U_i^g \equiv 0$. At the other extreme lies the purely hedonic case, with $U_i^c \equiv 0$ or $\ell = 0$, studied by Banerjee et al. (2001) and Bogomolnaia and Jackson (2002).

(PGE) implies absence of consumption externalities. But the latter does not imply the former because absence of consumption externalities does not necessarily mean that one can additively separate a pure consumption effect and a pure group effect. The effects of own private consumption and own group composition may be intertwined. For example, with $\ell \geq 2$, it could be that $U_i(\mathbf{x_h}; h) = \sum_{k=1}^{\ell} x_i^k$ if $h = \{i\}$ and $U_i(\mathbf{x_h}; h) = \prod_{k=1}^{\ell} x_i^k$ if $h \in \mathcal{H}_i$, $|h| > 1$. This example exhibits

(GSE) Group-Size Externalities: $U_i(\mathbf{x}; h) = V_i(x_i; |h|)$
for $i \in h, h \in \mathcal{H}, \mathbf{x} \in \mathcal{X}$.

In this case, individual i cares only about own consumption and household size. Still, preferences over own consumption may change with household size and, vice versa, preferences over household size can depend on own consumption. In the separable case, $U_i(\mathbf{x}; h) = u_i(x_i) + v_i(|h|)$, preferences over own consumption and preferences over household size are independent.

When (PGE) does not hold, separation with respect to the consumption of individual household members may be possible instead. Finally, as a polar case to individual separability, a consumer may only care about the aggregate consumption of his fellow household members. This gives rise to notions of local and global anonymity. Consumer preferences may also satisfy certain monotonicity properties. To formulate those and other properties, we introduce some more notation. Recall that for $i \in I$, $\mathcal{H}_i \equiv \{h \subseteq I | i \in H\}$. $\mathcal{H}_i$ denotes the set of potential households of which i would be a member. If $h \in \mathcal{H}_i$ and $\mathbf{x_h} \in \mathcal{X}_h$, then we can write $\mathbf{x_h} = (x_i, x_{h\backslash i})$ where $h\backslash i$ serves as shorthand for $h\backslash\{i\}$ and

$$x_{h\backslash i} \in X_{h\backslash i} = \prod_{j \in h\backslash i} X_j$$

describes the consumption of household members j other than i. For $x_{h\backslash i} \in X_{h\backslash i}$, $x_{h\backslash i} = (x_j)_{j \in h\backslash i}$ denote $\bar{x}_{h\backslash i} = \sum_{j \in h\backslash i} x_j$, the aggregate consumption of household members other than i. Now we are prepared to formulate certain externalities as well as separability and monotonicity properties. We commence with the latter.

(MON) Monotonicity: $U_i(x_i, x_{h\backslash i})$ is increasing in x_i
for all $i \in I, h \in \mathcal{H}_i$.
(SMO) Strict Monotonicity: $U_i(x_i, x_{h\backslash i})$ is strictly increasing in x_i
for all $i \in I, h \in \mathcal{H}_i$.
(NNE) Non-Negative Externalities: $U_i(x_i, x_{h\backslash i})$ is non-decreasing in $x_{h\backslash i}$
for all $i \in I, h \in \mathcal{H}_i$.
(NPE) Non-Positive Externalities: $U_i(x_i, x_{h\backslash i})$ is non-increasing in $x_{h\backslash i}$
for all $i \in I, h \in \mathcal{H}_i$.

(SEP) Separable Externalities: $U_i((x_i, x_{h\backslash i}); h) = u_i^h(x_i) + \sum_{j \in h\backslash i} v_{ij}^h(x_j)$
for $i \in I, h \in \mathcal{H}_i$.
(LAN) Local Anonymity: $U_i((x_i, x_{h\backslash i}); h) = U_i((x_i, \bar{x}_{h\backslash i}); h)$
for $i \in I, h \in \mathcal{H}_i$.
(GAN) Global Anonymity: $U_i((x_i, x_{h\backslash i}); h) = U_i((x_i, \bar{x}_{h\backslash i}))$ s
for $i \in I, h \in \mathcal{H}_i$.

Household Preferences. For a single-person household $h = \{i\}$, household preferences coincide with the preferences of consumer i on X_i. For a multi-member household h, household preferences reflect the preferences of its constituents. In general, they are represented by the preference relation $\succsim_h$ on $\mathcal{X}_h$, given by

$$\mathbf{x_h} \succsim_h \mathbf{y_h} \Longleftrightarrow [U_i(\mathbf{x_h}; h) \geq U_i(\mathbf{y_h}; h)\ \forall i \in h]$$

for $\mathbf{x_h}, \mathbf{y_h} \in \mathcal{X}_h$. The relation $\succsim_h$ is reflexive, transitive and, as a rule, incomplete. Consider instead a utilitarian social welfare function W_h for household h, that is $W_h : \mathcal{X}_h \to \mathbb{R}$ where $W_h(\mathbf{x_h}) = \sum_{i \in h} a_i\, U_i(\mathbf{x_h}; h)$ for $\mathbf{x_h} \in \mathcal{X}_h$ and $a_i > 0$ for $i \in h$. Then W_h defines a complete and transitive preference relation $\succsim_h^*$ on $\mathcal{X}_h$ that contains $\succsim_h$. Hence for any subset $\mathcal{C}_h$ of $\mathcal{X}_h$, the set $\arg\max_{\mathbf{x_h} \in \mathcal{C}_h} W_h(\mathbf{x_h})$ consists of vectors that are maximal within $\mathcal{C}_h$ with respect to $\succsim_h$. Instead of a utilitarian social welfare function, the household may employ another Paretian social welfare function. The household may have preferences that are complete, transitive and strictly monotone in the welfare of each member, but not representable by means of a utility function on $\mathcal{X}_h$. For example, let $h = \{1, 2\}$, $\ell = 2$, and let there be absence of externalities, specifically $U_i(\mathbf{x_h}; h) = x_i^1 x_i^2$ for $i = 1, 2$; $\mathbf{x_h} \in \mathcal{X}_h$. Moreover, let $\geq_2^{lex}$ denote the lexicographic order on $\mathbb{R}^2$. Define the preference relation $\succsim_h^{lex}$ on $\mathcal{X}_h$ by

$$\mathbf{x_h} \succsim_h^{lex} \mathbf{y_h} \Longleftrightarrow [(U_1(\mathbf{x_h}; h), U_2(\mathbf{x_h}; h)) \geq_2^{lex} (U_1(\mathbf{y_h}; h), U_2(\mathbf{y_h}; h))]$$

for $\mathbf{x_h}, \mathbf{y_h} \in \mathcal{X}_h$. Then $\succsim_h^{lex}$ has the asserted properties.

At times, it proves useful to consider strict preference by all household members. Formally, let the preference relation $\gg_h$ on $\mathcal{X}_h$ be given by

$$\mathbf{x_h} \gg_h \mathbf{y_h} \Longleftrightarrow [U_i(\mathbf{x_h}; h) > U_i(\mathbf{y_h}; h)\ \forall i \in h]$$

for $\mathbf{x_h}, \mathbf{y_h} \in \mathcal{X}_h$. The relation $\gg_h$ is irreflexive, transitive and incomplete. It is contained in $\succsim_h$.

Certain properties apply to households rather than individual members:

Definition (*Redistribution Property (RP)*). The Redistribution Property holds for household h if for any two bundles $\mathbf{x_h}, \mathbf{y_h} \in \mathcal{X}_h$ with $\mathbf{y_h} \succ_h \mathbf{x_h}$, there exists $\mathbf{z_h} \in \mathcal{X}_h$ such that $\sum_{i \in h} z_i = \sum_{i \in h} y_i$ and $\mathbf{z_h} \gg_h \mathbf{x_h}$.

At first sight, the Redistribution Property appears to be a weak constraint on the preferences of household members. However, (RP) can be violated when consumption externalities are strong and positive. In such circumstances, redistributing commodities from one household member to others may not improve the utility of the individuals receiving those consumption goods.

Definition A household h is **locally non-satiated** if for every $\mathbf{x_h} \in \mathcal{X}_h$ and every $\epsilon > 0$, there exists $\mathbf{y_h} \in \mathcal{X}_h$ with $\|\mathbf{x_h} - \mathbf{y_h}\|_{|h|\ell} < \epsilon$ and $\mathbf{y_h} \succ_h \mathbf{x_h}$ where $\|\cdot\|_d$ denotes the Euclidean norm on $\mathbb{R}^d$.

We note that (PR) and local non-satiation are independent properties. A household may be locally non-satiated while (RP) fails. Conversely, (RP) may hold, but the household may be locally satiated at some consumption bundle $\mathbf{x_h}$.

Household Decisions. The novelty of our general equilibrium analysis is the integration of three allocative mechanisms involving households: (a) Household formation (and dissolution), i.e., individual decisions are made to join or leave households. (b) Household decisions, i.e., collective decisions within households regarding the individual consumption plans of household members. (c) Competitive exchange across households.

Concerning (b), allocative decisions at the household level, the household makes collective decisions regarding the consumption of private goods by its members. These collective decisions rest on two premises. First, there is a joint budget constraint for all household members. Second, the household chooses an efficient consumption schedule for its members, subject to the household budget constraint.

Each consumer i belonging to household h has his own consumption set X_i and his own individual preferences represented by U_i. Our first premise says that the members of h are subject to a joint budget constraint. Formally, let us consider a household $h \in \mathcal{H}$ and a price system $p \in \mathbb{R}^\ell$. For $\mathbf{x_h} = (x_i)_{i \in h} \in \mathcal{X}_h$,

$$p * \mathbf{x_h} \equiv p \cdot \left(\sum_{i \in h} x_i \right)$$

denotes the expenditure of household h on household consumption plan $\mathbf{x_h}$ at the price system p. As p and $\mathbf{x_h}$ are of different dimension for multi-member households, we use the $*$-product in lieu of the familiar inner product. Then h's **budget set** is defined as

$$B_h(p) = \{\mathbf{x_h} \in \mathcal{X}_h \ : \ p * \mathbf{x_h} \leq p \cdot \omega_h\}.$$

Our second premise is motivated by Chiappori (1988, 1992) who, in contrast to traditional economic theory, suggests a model of collective rationality of households as an alternative to the neoclassical model where households are treated like single consumers. In its most general form, collective rationality à la Chiappori amounts to $\mathbf{x_h} \in EB_h(p)$ where the **efficient budget set** $EB_h(p)$ is defined as the set of $\mathbf{x_h} \in B_h(p)$ with the property that there is no $\mathbf{y_h} \in B_h(p)$ such that $\mathbf{y_h} \succ_h \mathbf{x_h}$, i.e., there is no $\mathbf{y_h} \in B_h(p)$ such that

$U_i(\mathbf{y_h}; h) \geq U_i(\mathbf{x_h}; h)$ for all $i \in h$;
$U_i(\mathbf{y_h}; h) > U_i(\mathbf{x_h}; h)$ for some $i \in h$.

In other words, the household chooses an element at the Pareto frontier of its budget set $B_h(p)$. Collective rationality that broadly defined leaves the exact decision criterion of the household unspecified. Different households may apply different criteria. For example, some households may maximize a Paretian social welfare function—with a Nash product as a special case. Others follow perhaps a rule that cannot be represented by an objective function for the household. For instance, with two goods, a two-person household $h = \{1, 2\}$ may pick a consumption plan that is a maximal element in $B_h(p)$ of the preference relation $\succsim_h^{lex}$ described earlier. We frequently work with utilitarian social welfare functions for households for the sake of examples or existence proofs. In such cases, household choice is not necessarily confined to the maximizers of the social welfare function at hand. For instance, there may exist competitive equilibria where the household's consumption plan does not maximize the particular social welfare function.

Our model of the household encompasses "unitary" models as special cases. Suppose that household h determines elements of $EB_h(p)$ by maximizing a utilitarian social welfare function W_h on $B_h(p)$. Then the household may be treated as a single consumer as far as the aggregate demand of the household is concerned. Namely, let $\mathcal{A}_h = \mathbb{R}^\ell_+$ and define a utility function $\mathfrak{U}_h : \mathcal{A}_h \to \mathbb{R}$ that reflects household preferences:

$$\mathfrak{U}_h(a_h) = \max_{\mathbf{x_h} \in \mathfrak{A}_h(a_h)} W_h(\mathbf{x_h})$$

where $\mathfrak{A}_h(a_h) = \{\mathbf{x_h} = (x_i)_{i \in h} \in \mathcal{X}_h | \sum_{i \in h} x_i = a_h\}$. Then $\mathbf{x_h} = (x_i)_{i \in h}$ maximizes W_h on $B_h(p)$ if and only if $\bar{a}_h = \sum_{i \in h} x_i$ maximizes $\mathfrak{U}_h$ on $\{a_h \in \mathcal{A}_h | pa_h \leq p\omega_h\}$. Hence the consumer with characteristics $(\mathcal{A}_h, \mathcal{U}_h, \omega_h)$ constitutes a representative consumer for the household. However, welfare and policy conclusions for the representative consumer need not equally hold for the individual consumers of an economy or a household. See Dow and Werlang (1988), Kirman (1992) and Jerison (2006).

We note that important contributions to the literature argue against assuming efficiency in household decision making (Lundberg and Pollak (2003), Konrad and Lommerud (1995, 2000)). For instance, Lundberg et al. (1997) provide evidence that tends to support the idea that household members do not pool their incomes as it would be implied by efficient collective decision making. But both the theoretical and the empirical literature appear to be split in this matter: Browning and Chiappori (1998, p. 1245) claim "support for our view that the collective model is a viable alternative to the unitary model." Browning et al. (2006, p. 6) list a number of different approaches to model intra-household bargaining. They further state that there is no broad consensus which particular model to use. For further discussion, we refer to Chap. 10 and Appendix 1 of Gersbach and Haller (2012).

References

Banerjee, S., Konishi, H., & Sönmez, T. (2001). Core in a simple coalition formation game. *Social Choice and Welfare*, *18*, 135–153.

Bogomolnaia, A., & Jackson, M. O. (2002). The stability of hedonic coalition structures. *Games and Economic Behavior*, *38*, 201–230.

Browning, M., & Chiappori, P.-A. (1998). Efficient intra-household allocations: A general characterisation and empirical tests. *Econometrica*, *66*, 1241–1278.

Browning, M., Chiappori, P.-A., & Lechene, V. (2006). Collective and unitary models: A clarification. *Review of Economics of the Household*, *4*, 5–14.

Chiappori, P.-A. (1988). Rational household labor supply. *Econometrica*, *56*, 63–89.

Chiappori, P.-A. (1992). Collective labor supply and welfare. *Journal of Political Economy*, *100*, 437–467.

Dow, J., & Werlang, S. (1988). The consistency of welfare judgments with a representative consumer. *Journal of Economic Theory*, *44*, 269–280.

Gersbach, H. & Haller, H. (2004). Hold-Up Problems and Firm Formation. CEPR Discussion Paper No. 4688.

Gersbach, H., & Haller, H. (2012). "Hard Workers" and labor restrictions. *Journal of Population Economics*, *25*, 469–494.

Gori, M., & Villanacci, A. (2011). A bargaining model in general equilibrium. *Economic Theory*, *46*, 327–375.

Jerison, M. (2006). Nonrepresentative Representative Consumers. Discussion Papers 06-08, SUNY, University at Albany, NY, Department of Economics.

Kirman, A. P. (1992). Whom or what does the representative individual represent? *Journal of Economic Perspectives*, *6*, 117–136.

Konrad, K., & Lommerud, K. E. (1995). Family policy with non-cooperative families. *Scandinavian Journal of Economics*, *97*, 581–601.

Konrad, K., & Lommerud, K. E. (2000). The bargaining family revisited. *Canadian Journal of Economics*, *33*, 471–487.

Lundberg, S., & Pollak, R. (2003). Efficiency in marriage. *Review of Economics of the Household*, *1*, 153–167.

Lundberg, S., Pollak, R., & Wales, T. (1997). Do husband and wives pool their resources? Evidence from the U.K. child benefit. *Journal of Human Resources*, *32*, 463–480.

Part II
The Basic Framework

Chapter 4
Pure Exchange with Fixed Household Structure

In this and the next three chapters we strive for the integration of the three previously highlighted allocative mechanisms involving households: (a) Household formation (and dissolution), i.e., individual decisions are made to join or leave households. (b) Household decisions, i.e., collective decisions within households regarding the individual consumption plans of household members. (c) Competitive exchange across households. Our goal is achieved in several stages which reflect the historical development. At the initial stage presented in the current chapter, the household structure is fixed. Hence only (b) household decisions and (c) competitive exchange across households are considered. The model analyzed in the next chapter incorporates production while the household structure is again fixed. Thereafter, we return to pure exchange economies. At the intermediate stage treated in Chap. 6, variable household structures are introduced. This innovation allows novel comparative statics and more elaborate welfare analysis. Still, the model is confined to (b) and (c). The final stage is presented in Chap. 7, where we add (a) household formation and dissolution.

To model (c) competitive exchange among multi-member households, we consider a pure exchange economy composed of finitely many households labeled $h = 1, \dots, H$. Throughout this chapter, the household structure $P = \{1, \dots, H\}$ is treated as exogenous (fixed). The commodity space is $\mathbb{R}^\ell$ with $\ell \geq 1$. Household $h \in P$ is endowed with a commodity bundle $\omega_h \in \mathbb{R}^\ell$, $\omega_h > 0$. The aggregate or social endowment is $\omega_P = \sum_{h \in P} \omega_h = \sum_{h=1}^{H} \omega_h$.

Each household h consists of finitely many members $i = hm$ with $m = 1, \dots, m(h)$ and $m(h) \geq 1$. Put $I = \{hm : h = 1, \dots, H; m = 1, \dots, m(h)\}$. A generic individual $i = hm \in I$ has:

- consumption set $X_i = \mathbb{R}^\ell_+$;
- preferences $\succsim_i$ on the allocation space $\mathcal{X} \equiv \prod_{j \in I} X_j$ represented by a utility function $U_i : \mathcal{X} \longrightarrow \mathbb{R}$.

H. Gersbach and H. Haller, *Groups and Markets*,
DOI 10.1007/978-3-319-60516-6_4

This formulation allows for economy-wide externalities. The latter promises to be a fertile topic of research even in the traditional context of competitive exchange among individuals. But in accordance with the topic of the book, we propose to restrict attention to externalities that are of particular interest for an inquiry into competitive exchange among households. Hence in the sequel, condition (E1) will be imposed which requires that consumption externalities, if any, exist only between members of the same household. Some more notation is needed for an explicit formulation of such intra-household externalities.

Let $\mathbf{x}=(x_i)$, $\mathbf{y}=(y_i)$, $\mathbf{z}=(z_i)$ denote generic elements of $\mathcal{X}$. For $h = 1, \ldots, H$, define $\mathcal{X}_h = \prod_{n=1}^{m(h)} X_{hn}$ with generic elements $\mathbf{x_h} = (x_{h1}, \ldots, x_{hm(h)})$. If $\mathbf{x} \in \mathcal{X}$ is an allocation, then for $h = 1, \ldots, H$, household consumption is $\mathbf{x_h} = (x_{h1}, \ldots, x_{hm(h)}) \in \mathcal{X}_h$. Now we are ready to define the kind of intra-household externalities which will be assumed hereafter.

(E1) Intra-Household Externalities (Household-Specific Preferences):

$$U_i(\mathbf{x}) = U_i(\mathbf{x_h}) \quad \text{for} \quad i = hm, \ \mathbf{x} \in \mathcal{X}.$$

(E1) is obtained from (HSP) when the household structure is fixed and consumer hm's household affiliation, h is omitted as an argument of the utility function. We shall also refer to the special case of no externalities, i.e.,

(E2) Absence of Externalities:

$$U_i(\mathbf{x}) = u_i(x_i) \text{ for } i = hm, \mathbf{x} = (x_i) \in \mathcal{X}.$$

The first theorem of welfare economics asserts that any competitive equilibrium allocation in the sense of Walras is Pareto-optimal—with the usual qualification of local non-satiation. Here, like in Haller (2000), we want to allow for the possibility of a household composed of several members who arrive at a collective decision on household consumption. For the economy with social endowment $\omega_P = \sum_h \omega_h$ and consumers $i = hm$ ($h = 1, \ldots, H; m = 1, \ldots, m(h)$), an efficient or Pareto-optimal allocation is defined in the standard fashion based on individual preferences:

Definition An allocation $\mathbf{x} = (x_i) \in \mathcal{X}$ is called ***P*-efficient**, ***P*-optimal** or **constrained Pareto-optimal (given *P*)**, if

(i) $\sum_i x_i = \omega_P$, i.e., $\mathbf{x}$ is P-feasible and
(ii) there does not exist a P-feasible allocation $\mathbf{y} = (y_i) \in \mathcal{X}$ with $U_i(\mathbf{y}) \geq U_i(\mathbf{x})$ for all i and $U_i(\mathbf{y}) > U_i(\mathbf{x})$ for some i.

Let us denote the P-feasible allocations defined via (i) by $\mathcal{X}(P)$ and the P-optimal allocations by $\mathcal{M}(P)$. The notation $\mathcal{M}(P)$ suggests that $\mathcal{M}(P)$ consists of the maximal elements with respect to the "Pareto ordering" $\succ_P$ on $\mathcal{X}(P)$, defined by $\mathbf{y} \succ_P \mathbf{x} \Longleftrightarrow (U_i(\mathbf{y}))_{i\in I} > (U_i(\mathbf{x}))_{i\in I}$ for $\mathbf{x}, \mathbf{y} \in \mathcal{X}(P)$. We obtain:

Proposition 4.1 (Gersbach and Haller (2001), Proposition 1)
Suppose all utility functions $U_i(\cdot;\ h)$, $i \in I, h \in \mathcal{H}_i \cap P$, are continuous. Then:

(i) $\mathcal{M}(P) \neq \emptyset$.
(ii) For each $\mathbf{y} \in \mathcal{X}(P)$, *there exists* $\mathbf{x} \in \mathcal{M}(P)$ *with* $(U_i(\mathbf{x}))_{i\in I} \geq (U_i(\mathbf{y}))_{i\in I}$.

To complete the modeling of competitive exchange among households, one has to specify how households interact with the market. Haller (2000) assumes efficient bargaining within households. The latter means that a household h chooses an allocation at the Pareto frontier of its budget set, i.e., an element of its efficient budget set $EB_h(p)$ as defined below. In contrast, Gersbach and Haller (2005, 2006) are investigating the impact of inefficient household decisions on market performance. This extended research agenda necessitates a more general definition of a competitive equilibrium among households than the one adopted in Haller (2000). To this end, consider a household h and a price system $p \in \mathbb{R}^\ell$. For $\mathbf{x_h} = (x_{h1}, \ldots, x_{hm(h)}) \in \mathcal{X}_h$, let

$$p * \mathbf{x_h} = p \cdot \left(\sum_{m=1}^{m(h)} x_{hm} \right)$$

denote the expenditure of household h on household consumption plan $\mathbf{x_h}$ at the price system p. As p and $\mathbf{x_h}$ are of different dimension for multi-member households, we use the $*$-product in lieu of the familiar inner product. Then h's **budget set** is defined as $B_h(p) = \{\mathbf{x_h} \in \mathcal{X}_h \ : \ p * \mathbf{x_h} \leq p \cdot \omega_h\}$.

For future reference, we also define household h's **binding budget set** or **balanced budget set** as $BB_h(p) = \{\mathbf{x_h} \in \mathcal{X}_h \ : \ p * \mathbf{x_h} = p \cdot \omega_h\}$.

Before we turn to the leading equilibrium concept involving efficient household decisions, let us digress a bit to point out the potential scope of general equilibrium analysis with multi-member households. In the broader context, demand correspondences describe the possible outcomes of collective household decision making. A correspondence

$$D_h : \mathbb{R}^\ell \twoheadrightarrow \mathcal{X}_h$$

is called a **demand correspondence for household** h, if $D_h(p) \subseteq B_h(p)$ for all $p \in \mathbb{R}^\ell$. How households form their demands is a key component of the definition of a competitive equilibrium among households. Given a profile $D = (D_1, \ldots, D_H)$ of demand correspondences for households, we can define a **competitive equilibrium with respect to the profile** D as a pair $(p; \mathbf{x})$ consisting of a price system p and a feasible commodity allocation $\mathbf{x} = (x_i)$ such that $\mathbf{x_h} \in D_h(p)$ for $h = 1, \ldots, H$. Thus in a competitive equilibrium, each household makes a collective choice under its budget constraint and markets clear. At this general level, the concept of a competitive equilibrium among households is flexible enough to accommodate all conceivable collective decision criteria of households. Of course, additional restrictions on the profile D could and should be imposed whenever warranted by the objective of the research effort. In Gersbach and Haller (2005, 2006), we introduced this general formulation to accommodate inefficient household decisions whereas in most of our work we impose the restriction that households make efficient choices under their budget constraints.

4.1 Efficient Household Decisions

Efficient choice by the household refers to the individual consumption and welfare of its members, not merely to the aggregate consumption bundle of the household. Such a notion of efficient household decision is captured by the concept of an efficient budget set.

Given a price system p, define household h's **efficient budget set** $EB_h(p)$ as the set of $\mathbf{x_h} \in B_h(p)$ with the property that there is no $\mathbf{y_h} \in B_h(p)$ such that

$$U_{hm}(\mathbf{y_h}) \geq U_{hm}(\mathbf{x_h}) \text{ for all } m = 1, \ldots, m(h);$$
$$U_{hm}(\mathbf{y_h}) > U_{hm}(\mathbf{x_h}) \text{ for some } m = 1, \ldots, m(h).$$

Classical versions of the first theorem of welfare economics are based on the crucial property that each consumer's demand lies on the consumer's budget line or hyperplane—which implies Walras' Law. This property follows from local non-satiation, for instance monotonicity of consumer preferences. With the possibility of multi-person households and intra-household externalities, the crucial property needs to be adapted. The modified property is called budget exhaustion and stipulates that each household's choice lies on the household's "budget line". For example, monotonicity in own consumption combined with non-negative externalities yields budget exhaustion. The formal definition is as follows.

(BE) Budget Exhaustion: For each household $h = 1, \ldots, H$, and any price system $p \in \mathbb{R}^\ell$, $EB_h(p) \subseteq BB_h(p)$.

Notice that $EB \equiv (EB_1(\cdot), \ldots, EB_H(\cdot))$ is an example of a profile of demand correspondences for households. To be consistent with the notation and terminology in other chapters, we call a competitive equilibrium with respect to the profile EB a P-equilibrium. This convention indicates that the household structure P prevails. Formally:

Definition A pair $(p; \mathbf{x})$ is a **competitive equilibrium among households (given household structure P)** or a **P-equilibrium** if p is a price system and $\mathbf{x} = (x_i)$ is a P-feasible commodity allocation such that

(iii) $\mathbf{x_h} \in EB_h(p)$ for all $h \in P$.

Condition (iii) is the formal expression of "collective rationality" of households. A key result of Haller (2000) can be rephrased as follows.

Proposition 4.2 (First Welfare Theorem) *Suppose (E1) and (BE).*
If $(p; \mathbf{x})$ *is a* P*-equilibrium, then* $\mathbf{x} \in \mathcal{M}(P)$.

In other words, efficiency at the household level implies efficiency at the economy level, if each household has to exhaust its budget in order to put into effect an efficient consumption decision for its members. Without (BE), P-equilibrium allocations are still weakly Pareto-optimal, but not necessarily Pareto-optimal in the strong sense of P-optimality. This fact is illustrated by Example 3.3 of Haller (2000):

Let $\ell = 1$, $H = 2$ and the two households be $h = \{1, 2\}$, $k = \{3\}$. Their endowments are $\omega_h = 2$ and $\omega_k = 1$. The utility functions of consumers are $U_1(x_1, x_2) = x_1 - x_2$, $U_2(x_1, x_2) = x_2 - x_1$, $U_3(x_3) = x_3$.

$p^* = 1$ and $\mathbf{x}^* = (1, 1, 1)$ constitute a P-equilibrium. There is no feasible allocation that strictly Pareto-dominates $\mathbf{x}^*$. Hence $\mathbf{x}^*$ is weakly Pareto-optimal. However, $\mathbf{x}^*$ is weakly Pareto-dominated by the feasible allocation $\mathbf{y} = (1/2, 1/2, 2)$. Therefore, $\mathbf{x}^*$ is not strongly Pareto-optimal. □□

Haller (2000) goes on and defines the "household core". He finds that under (E1) and (BE), P-equilibrium allocations belong to the household core. See Proposition 3 in Haller (2000) and Proposition 8.1 below. (BE) proves also conducive to the existence of P-equilibria.

Second Welfare Theorem. The welfare analysis of Haller (2000) is confined to the first theorem of welfare economics (and household core inclusion). But further conclusions obtain: The second welfare theorem in Gersbach and Haller (2001) (Proposition 6.4 below) immediately applies here.

4.2 Equilibrium Existence

There are several approaches to show existence of P-equilibria, pursued by Sato (2009), Gori (2010) and ourselves.

Proposition 4.3 (Sato (2009), Theorem 1)
Suppose (BE) and for all $h \in P$,

(i) $\omega_h \gg 0$;
(ii) U_{hm} *is continuous and concave on* $\mathcal{X}_h$ *for all* $m = 1, \ldots, m(h)$.

Then, there exists a P*-equilibrium* $(p^*; \mathbf{x}^*)$ *with* $p^* \neq 0$.

Sato's existence theorem is based on a result for polar cones in Geistdoerfer-Florenzano (1982). It allows for $p^* \ngtr 0$. Gersbach and Haller (1999) show a different result under similar assumptions. Their proof actually assumes "budget exhaustion for truncated budget sets", formulated as (BE') hereafter. Assuming $\omega_h \gg 0$ for all households h, consider hypercubes of the form $K = [0, k]^\ell$ such that ω_h belongs to the interior of K^h for all h. Define truncated budget sets $B_h^k(p) = B_h(p) \cap K^h$ and corresponding efficient truncated budget sets $EB_h^k(p)$, replacing $B_h(p)$ by $B_h^k(p)$ in the definition of the latter. Then "budget exhaustion for truncated budget sets" amounts to

(BE') Truncated Budget Exhaustion: For each household $h \in P = \{1, \ldots, H\}$, any price system $p \in \mathbb{R}^\ell$ and sufficiently large k, $EB_h^k(p) \subseteq BB_h(p)$.

Proposition 4.4 (Gersbach and Haller (1999), Proposition 1)
Suppose (BE') and for all $h \in P$,

(i) $\omega_h \gg 0$;
(ii) U_{hm} *is continuous and concave on* $\mathcal{X}_h$ *for all* $m = 1, \ldots, m(h)$.

Then, there exists a P-equilibrium $(p^*; \mathbf{x}^*)$ *with free disposal and* $p^* > 0$.

Free disposal allows for allocations $\mathbf{x} = (x_i) \in \mathcal{X}$ that satisfy $\sum_i x_i \leq \omega_P$ or aggregate excess demand $z = \sum_i x_i - \omega_P$ satisfying $z \leq 0$ in contrast to the standard feasibility (exact market clearing) condition $\sum_i x_i = \omega_P$ or $z = 0$. It turns out that (BE') implies (BE), but not vice versa:

First, suppose (BE) is violated. Then there exist a household h, a price system $p \in \mathbb{R}^\ell$ and $\mathbf{x_h} \in EB_h(p)$ such that $p * \mathbf{x_h} < p \cdot \omega_h$. For sufficiently large k, $\mathbf{x_h} \in B_h^k(p)$. Since $\mathbf{x_h} \in EB_h(p)$, there is no $\mathbf{y_h} \in B_h(p)$ such that $U_{hm}(\mathbf{y_h}) \geq U_{hm}(\mathbf{x_h})$ for all $m = 1, \ldots, m(h)$ and $U_{hm}(\mathbf{y_h}) > U_{hm}(\mathbf{x_h})$ for some $m = 1, \ldots, m(h)$. Because $B_h^k(p) \subseteq B_h(p)$, it follows that there is no $\mathbf{y_h} \in B_h^k(p)$ such that $U_{hm}(\mathbf{y_h}) \geq U_{hm}(\mathbf{x_h})$ for all $m = 1, \ldots, m(h)$ and $U_{hm}(\mathbf{y_h}) > U_{hm}(\mathbf{x_h})$ for some $m = 1, \ldots, m(h)$. This combined with $\mathbf{x_h} \in B_h^k(p)$ means that $\mathbf{x_h} \in EB_h^k(p)$ whereas $p * \mathbf{x_h} < p \cdot \omega_h$. Thus (BE') is violated. We have shown $\neg(BE) \Rightarrow \neg(BE')$ and, therefore, $(BE') \Rightarrow (BE)$.

Second, consider the case $\ell = 2$ and the only household h consisting of a single consumer i, with endowment bundle $\omega_{\{i\}} = (1, 1)$ and utility representation $U_i(x_i^1, x_i^2) = x_i^1$ for $(x_i^1, x_i^2) \in \mathbb{R}_+^2$. For all $p \in \mathbb{R}^2$ and $(x_i^1, x_i^2) \in \mathbb{R}_+^2$ with $p \cdot (x_i^1, x_i^2) < p \cdot \omega_{\{i\}}$, there exists $\varepsilon > 0$ such that $p \cdot (x_i^1 + \varepsilon, x_i^2) < p \cdot \omega_{\{i\}}$ and $U_i(x_i^1 + \varepsilon, x_i^2) > U_i(x_i^1, x_i^2)$. Hence $(x_i^1, x_i^2) \notin BB_h(p) \Rightarrow (x_i^1, x_i^2) \notin EB_h(p)$ and, consequently, $(x_i^1, x_i^2) \in EB_h(p) \Rightarrow (x_i^1, x_i^2) \in BB_h(p)$. This shows (BE). Next let $k = 2$ and $p = (1, 2)$. Then $(x_i^1, x_i^2) = (2, 0) \in EB_h^k(p)$ while $p * (x_i^1, x_i^2) = 2$ and $p \cdot \omega_{\{i\}} = 3$. Thus $(x_i^1, x_i^2) \in EB_h^k(p)$ and $(x_i^1, x_i^2) \notin BB_h(p)$, a violation of (BE'). This shows $(BE) \nRightarrow (BE')$.

One arrives at further conclusions with the following assumption for all or some consumers i:

(M_i) U_i is strictly monotonic and exhibits non-negative externalities.

Corollary 4.1 *If in Proposition 4.3 (or Proposition 4.4), assumption* (M_i) *is added for all members i of some household h, then the equilibrium price system satisfies* $p^* \gg 0$ *and exact market clearing obtains.*

We need not assume $\omega_h \gg 0$ for all households h if we assume (M_i) for all consumers in a household. This is the case in the following two propositions:

Proposition 4.5 (Gersbach and Haller (1999), Proposition 3) *Suppose that $\omega_P \gg 0$ and for each consumer i, (M_i) holds and U_i is continuous and strictly concave. Then a P-equilibrium $(p^*; \mathbf{x}^*)$ with exact market clearing and $p^* \gg 0$ exists.*

Proposition 4.6 (Gersbach and Haller (1999), Proposition 4) *Suppose that U_i is continuous and strictly concave for each consumer i. Further assume that at least one household h satisfies $\omega_h \gg 0$ and (M_i) for all $i \in h$. Then a P-equilibrium $(p^*; \mathbf{x}^*)$ with free disposal and $p^* \gg 0$ exists.*

Modification of standard arguments reduces the proof of both propositions to the excess demand lemma, which is a consequence of Brouwer's fixed point theorem. Let $\mathring{\Delta}$ denote the relative interior and $\partial\Delta$ denote the boundary of the unit price simplex Δ. For $z : \mathring{\Delta} \to \mathbb{R}^\ell$, we consider the following conditions:

Strong Walras' Law: $p \cdot z(p) = 0$ for all $z \in \mathring{\Delta}$.
Continuity Condition: z is continuous and bounded from below.
Boundary Condition: If $p \in \partial\Delta$ and $(p_n)_{n\in\mathbb{N}}$ is a sequence in $\mathring{\Delta}$ with $p_n \to p$, then $\|z(p_n)\|_\ell \to \infty$, where $\|\cdot\|_\ell$ denotes the Euclidean norm on $\mathbb{R}^\ell$.

Lemma 1 (Excess Demand Lemma)
Let $\ell \geq 2$.
Suppose $z : \mathring{\Delta} \to \mathbb{R}^\ell$ satisfies the Strong Walras' Law and the above continuity and boundary conditions. Then there exists $p^* \in \mathring{\Delta}$ with $z(p^*) = 0$.

See Proposition 17.C.1 in Mas-Collell et al. (1995). Under the hypothesis of Proposition 4.5, the aggregate excess demand has the desired properties. To show this claim, adapt the proof of Lemma 4 in Debreu (1982) and the proof of Proposition 17.B.2 in Mas-Collell et al. (1995). Under the hypothesis of Proposition 4.6, the excess demand of some household satisfies the boundary condition. The excess demand $z_h(p)$ of any household h satisfies $p \cdot z_h(p) \leq 0$ for all $p \in \mathring{\Delta}$ and the continuity condition. Replace then $z_h(p)$ by the artificial excess demand $\bar{z}_h(p) = z_h(p) + \frac{|p \cdot z_h(p)|}{\|p\|_\ell^2} p$. Then the aggregate artificial excess demand has the desired properties.

Sato (2009) derives another existence result without assuming budget exhaustion. For a household h, he defines (in our notation) $\widehat{\mathcal{X}}_h = \{\mathbf{x_h} \in \mathcal{X}_h | \sum_m x_{mh} \leq \omega_S\}$. He calls a consumer mh locally non-satiated on $\widehat{\mathcal{X}}_h$ if for all $\mathbf{x_h} \in \widehat{\mathcal{X}}_h$ and $\varepsilon > 0$, there exists $\mathbf{y_h} \in \mathcal{X}_h$ such that $\|\mathbf{y_h} - \mathbf{x_h}\|_{\ell \cdot m(h)} < \varepsilon$ and $U_{mh}(\mathbf{y_h}) > U_{mh}(\mathbf{x_h})$. He shows existence of a P-equilibrium if in Proposition 4.3, (BE) is replaced by the assumption that each household h has a member who is locally non-satiated on $\widehat{\mathcal{X}}_h$. Gori (2010), using homotopy techniques, obtains equilibrium existence for economies with Nash-bargained household decisions and no intra-household consumption externalities. In Chap. 5, equilibrium existence for a model with multi-member households and production is shown by following a strategy of proof different from Gersbach and Haller

(1999) and Sato (2009) on the one hand and Gori (2010) on the other hand. There, we define induced household preferences for aggregate household consumption and consider an artificial economy where household choice is aggregate consumption so that household choices and production plans reside in the same Euclidean space. Resorting to a theorem of Debreu (1982) based on the simultaneous optimization or social equilibrium approach of Arrow and Debreu (1954), we then show existence of a competitive equilibrium for this artificial economy—which translates into an equilibrium of the actual economy.

4.3 Ramifications

Several interesting model variations are only briefly mentioned here and discussed in more detail in other chapters.

Individual Decentralization. An alternative to the unitary model of the household on the one hand and the collective rationality model on the other hand is the use of a budget sharing rule. As we elaborate in Chap. 13, any collectively rational decision can be reached with a suitable sharing rule if there are no consumption externalities (and preferences are locally non-satiated). As a rule, however, with positive or negative consumption externalities, a household cannot achieve collectively rational outcomes by simply applying a budget sharing rule and allowing household members shop on their own.

Inefficient Household Decisions. Household members may be unwilling or unable to make an efficient household decision or households may just make mistakes. In any case, inefficient household decisions can result. In Gersbach and Haller (2005) we embark on an alternative research agenda and study the causes and effects of inefficient household decisions. The main findings are highlighted in Chap. 13.

Comparative Statics. The restriction $\mathbf{x_h} \in EB_h(p)$ does not presume any numeric objective function for household h, but it does not rule out maximization of a specific objective function either. Instances of the latter are maximization of a Nash product or a utilitarian social welfare function. The respective bargaining weights or welfare weights may be exogenous or endogenous. A change of exogenous weight parameters can be viewed as shift of formal power within the household and render intriguing comparative statics. Details of the latter can be found in Gersbach and Haller (2009, 2012) and a summary in Chap. 11.

Production. In Chap. 5, we take a first pass at a general equilibrium model with multi-member households and production. We address the question whether existence results and the first welfare theorem can be extended from a pure exchange context to a model with production.

References

Arrow, K. J., & Debreu, G. (1954). Existence of an equilibrium for a competitive economy. *Econometrica, 22*, 265–290.

Debreu, G. (1982). Existence of competitive equilibrium. In K. J. Arrow & M. D. Intriligator (Eds.), *Handbook of Mathematical Economics* (Vol. 2). Amsterdam: North-Holland.

Geistdoerfer-Florenzano, M. (1982). The Gale-Nikaido-Debreu lemma and the existence of transitive equilibrium with or without the free-disposal assumption. *Journal of Mathematical Economics, 9*, 113–134.

Gersbach, H., & Haller, H. (1999). Allocation among multi-member households: Issues, cores and equilibria. In A. Alkan, C. D. Aliprantis, & N. C. Yannelis (Eds.), *Current trends in economics: Theory and applications*. Berlin, Heidelberg: Springer-Verlag.

Gersbach, H., & Haller, H. (2001). Collective decisions and competitive markets. *Review of Economic Studies, 68*, 347–368.

Gersbach, H., & Haller, H. (2005). When inefficiency begets efficiency. *Economic Theory, 25*, 105–121.

Gersbach, H., & Haller, H. (2006). Household inefficiency and equilibrium efficiency. In C. Schultz & K. Vind (Eds.), *Institutions, equilibria and efficiency: Essays in honor of Birgit Grodal* (pp. 187–209). Heidelberg: Springer-Verlag.

Gersbach, H., & Haller, H. (2009). Bargaining power and equilibrium consumption. *Social Choice and Welfare, 33*, 665–690.

Gersbach, H., & Haller, H. (2012). "Hard workers" and labor restrictions. *Journal of Population Economics, 25*, 469–494.

Gori, M. (2010). Endogenous household formation and inefficiency in a general equilibrium model. Working Paper, Dipartimento di Matematica per le Decisioni, Università degli Studi di Firenze.

Haller, H. (2000). Household decisions and equilibrium efficiency. *International Economic Review, 41*, 835–847.

Mas-Collell, A., Whinston, M. D., & Green, J. R. (1995). *Microeconomic theory*. Oxford, UK: Oxford University Press.

Sato, N. (2009). Existence of competitive equilibrium in economies with multi-member households. *Economics Bulletin, 29*, 1760–1771.

Chapter 5
General Equilibrium with Fixed Household Structure and Production

In this chapter, we consider firms and multi-member households operating in a competitive market environment. Households are endowed with resources (commodity bundles) and shares of firm ownership. Household members are characterized by individual preferences, possibly with intra-household consumption externalities. Household decisions adhere to the collective rationality model. Existence of general equilibrium and validity of the first welfare theorem are investigated.

5.1 Introduction

Traditional economic theory has treated "households" and "consumers" as synonyms. But would a formal distinction between a household as an economic entity and its constituents make a significant difference beyond a mere descriptive improvement? It does, if one is interested in household labor supply to market as well as household production, in the differential effect of taxes, subsidies and public goods on household members, to name just a few instances. Numerous theoretical and empirical studies have examined household related issues, usually relying on partial equilibrium analysis. Here like throughout this book, we are interested in the behavior and welfare of multi-member households in a general equilibrium context. This framework allows to investigate the feedback between decisions at the micro-level, by households and their members, and macro-variables, in particular market clearing prices. In special cases, one is able to perform comparative statics with respect to exogenous model parameters, like in Gersbach and Haller (2009, 2012). Incorporating multi-member households requires specifying the decision making of such households. Haller (2000) pioneered the analysis of general equilibrium models with multi-member households operating in a competitive market environment. His approach was motivated and influenced by the model of collective rationality of households

H. Gersbach and H. Haller, *Groups and Markets*,
DOI 10.1007/978-3-319-60516-6_5

forwarded by Chiappori (1988, 1992).[1] Haller, whose approach is outlined in the previous chapter, considers a finite pure exchange economy and assumes collective rationality in its most general form: A household acts collectively in the market, with efficient bargaining within the household. In a competitive equilibrium, each household makes an efficient choice under its budget constraint, and markets clear. The crucial feature is that efficient choice by the household refers to the individual consumption and welfare of its members, not merely to the aggregate consumption bundle of the household.

First and foremost, two questions arise once a general equilibrium model with multi-member households is developed: Does the presence of multi-member households impair the efficiency of competitive equilibrium allocations where efficiency or Pareto optimality, to be precise, is defined in terms of individual preferences? Does the presence of multi-member households impede the existence of competitive equilibria? *Prima facie*, one might be inclined to think that the welfare properties of competitive equilibrium allocations depend on the details of intra-household bargaining. The key insight of Haller (2000) is that those details do not matter for the validity of the first welfare theorem. It suffices that in equilibrium, every household makes an efficient choice under its budget constraint and, by doing so, exhausts its budget. This neither requires nor rules out specific bargaining protocols or decision rules as long as efficient household decisions are reached—and the budget gets exhausted. Regarding the second question, it turns out that the aggregate excess demand of a multi-member household has similar properties as the excess demand of traditional consumers, which suggests that an equilibrium existence result should obtain via the excess demand approach. Gersbach and Haller (1999) and Sato (2009) take the excess demand approach to show equilibrium existence for economies à la Haller (2000). Gori (2010), using homotopy techniques, obtains equilibrium existence for economies with Nash-bargained household decisions and no intra-household consumption externalities. For further details, see Sect. 4.2.

The existing body of work on general equilibrium models with multi-member households has been confined to pure exchange economies.[2] Here we take a first pass at a general equilibrium model with multi-member households and production. We address the question whether equilibrium existence results and the first welfare theorem can be extended from a pure exchange context to a model with production. Extending results for pure exchange economies to economies with production is often, but not always straightforward. A more or less straightforward extension holds true for equilibrium existence and the first and second welfare theorem for finite Arrow-Debreu economies, though the proofs turn out to be more elaborate for economies with production than for pure exchange economies.

We find that extension of the first welfare theorem proves rather straightforward, indeed, though one of us—like perhaps the more sceptical or more cautious reader—

---

[1]Alternative models of household decision making have been introduced by Lundberg and Pollak (1993, 1994) and Apps and Rees (2009), among others.

[2]See Gersbach and Haller (1999, 2001, 2005, 2009, 2010, 2011), Gori (2010), Gori and Villanacci (2011), Haller (2000).

had to be convinced by an explicit proof. Equilibrium existence with production is shown by following a strategy of proof different from Gersbach and Haller (1999) and Sato (2009) on the one hand and Gori (2010) on the other hand. We define induced household preferences for aggregate household consumption so that household choices and production plans reside in the same Euclidean space. Resorting to a theorem of Debreu (1982) based on the simultaneous optimization or social equilibrium approach of Arrow and Debreu (1954), we then show existence of a competitive equilibrium for this artificial economy—which translates into an equilibrium of the actual economy.

In the next section, we introduce the model and state and demonstrate our first main result, a first welfare theorem for finite economies with multi-member households and production. Section 5.3 is devoted to equilibrium existence. In Sect. 5.4, we briefly discuss several ramifications. Section 5.5 offers concluding remarks.

5.2 Model and First Main Result

We consider an economy with a finite number of commodities, firms and households. The main departure from the traditional model is that a household can have several members, each with their own preferences. There are $\ell \geq 1$ continuous commodities, labeled $l \in \{1, \ldots, \ell\}$. Thus the commodity space is $\mathbb{R}^{\ell}$.

The population of consumers is divided into finitely many households $h = 1, \ldots, n$, with $n \geq 2$. Each household h consists of finitely many members $i = hm$ with $m = 1, \ldots, m_h$, $m_h \geq 1$. Put $I = \{hm : h = 1, \ldots, n;\ m = 1, \ldots, m_h\}$, the finite population of individuals to be considered. There are finitely many firms $j = 1, \ldots, f$, with $f \geq 1$. Let $H = \{1, \ldots, n\}$ denote the set of households and $J = \{1, \ldots, f\}$ denote the set of firms.

5.2.1 Technologies and Firm Decisions

Each firm j has a non-empty production set or technology $Y_j \subseteq \mathbb{R}^{\ell}$. The special case of $Y_j = \{0\}$ for all j amounts to a pure exchange economy. The objective of a firm is to maximize its profit, to the extent possible. For a price system $p \in \mathbb{R}^{\ell}$ and a firm j, let

$$Y_j(p) = \arg\max_{y_j \in Y_j} p y_j$$

be the set of j's profit maximizers. $Y_j(p)$ may be empty. In case $Y_j(p) \neq \emptyset$, set

$$\pi_j(p) = \max_{y_j \in Y_j} p y_j,$$

so that $\pi_j(p) = p y_j$ for $y_j \in Y_j(p)$.

5.2.2 Allocations and Individual Preferences

A generic individual $i = hm \in I$ has consumption set $X_i = \mathbb{R}^\ell_+$. Let $\mathcal{X} \equiv \prod_{i \in I} X_i$ be the set of consumption profiles and $\mathcal{Y} \equiv \prod_{j \in J} Y_j$ be the set of production profiles. Then the allocation space is $\mathcal{X} \times \mathcal{Y}$.

The consumption bundle of a generic individual i is denoted by x_i with $x_i \in X_i$. Let $\mathbf{x} = (x_i)$, $\mathbf{x}' = (x_i')$ denote generic elements of $\mathcal{X}$. For $h = 1, \ldots, n$, define $\mathcal{X}_h = \prod_{m=1}^{m_h} X_{hm}$ with generic elements $\mathbf{x_h} = (x_{h1}, \ldots, x_{hm_h})$. If $\mathbf{x} \in \mathcal{X}$ is a consumption profile, then for $h = 1, \ldots, n$, household consumption is given by $\mathbf{x_h} = (x_{h1}, \ldots, x_{hm_h}) \in \mathcal{X}_h$. We will allow for the possibility of consumption externalities. Following Haller (2000), we shall restrict attention to the case where such consumption externalities, if any, exist only between members of the same household. This is captured by the notion of **intra-household externalities**: For $i \in h$, the welfare of individual i depends only on household consumption $\mathbf{x_h}$. More specifically, we assume that the preferences of individual i have a **utility representation** $U_i : \mathcal{X}_h \longrightarrow \mathbb{R}$.

We adopt from Gersbach and Haller (2001) the concept of local non-satiation of multi-person households:

Definition A household h is **locally non-satiated** if for every $\mathbf{x_h} \in \mathcal{X}_h$ and every $\epsilon > 0$, there exists $\mathbf{x_h'} \in \mathcal{X}_h$ with

$$\| \mathbf{x_h} - \mathbf{x_h'} \|_{m_h \ell} < \epsilon \text{ and } (U_i(\mathbf{x_h'}))_{i \in h} > (U_i(\mathbf{x_h}))_{i \in h}.$$ [3]

For local non-satiation of household h to hold it suffices that there exist a member hm and a commodity $c(h) \in \{1, \ldots, \ell\}$ such that (a) the welfare of hm is strictly increasing in $x_{hm}^{c(h)}$, hm's consumption of commodity $c(h)$ and (b) the welfare of all other household members is unaffected or positively affected by hm's consumption of commodity $c(h)$.

Local non-satiation for all households implies the budget exhaustion property (5.2) assumed in Proposition 5.1. Local non-satiation for all households is also one of the assumptions of Proposition 5.2.

5.2.3 Property Rights and Household Decisions

Household h is endowed with a commodity bundle $\omega_h \in \mathbb{R}^\ell$, $\omega_h > 0$. The aggregate or social endowment is $\omega = \sum_h \omega_h$. Moreover, household h owns a share $\theta_{hj} \geq 0$ of firm $j \in J$. For each firm, $\sum_h \theta_{hj} = 1$.

Given a price system $p \in \mathbb{R}^\ell$ such that $Y_j(p) \neq \emptyset$ for all j, household h has wealth or income

[3] $\| \cdot \|_d$ denotes the Euclidean norm on a d-dimensional Euclidean space. We use the notation $\gg$, $>$ and $\geq$ for vector inequalities.

$$\mathrm{w}_h(p) = p\,\omega_h + \sum_{j \in J} \theta_{hj}\pi_j(p).$$

Now consider a household h and a price system $p \in \mathbb{R}^\ell$. For $\mathbf{x_h} = (x_{h1}, \ldots, x_{hm_h}) \in \mathcal{X}_h$,

$$p * \mathbf{x_h} = p \cdot \left(\sum_{m=1}^{m_h} x_{hm} \right)$$

denotes the total household expenditure on household consumption plan $\mathbf{x_h}$ at the price system p. As p and $\mathbf{x_h}$ are of different dimension for multi-member households, we use the $*$-product in lieu of the familiar inner product. If household wealth or income $\mathrm{w}_h(p)$ is well defined, then h's **budget set** is given as $B_h(p) = \{\mathbf{x_h} \in \mathcal{X}_h : p * \mathbf{x_h} \le \mathrm{w}_h(p)\}$. Next we define the **efficient budget set** $EB_h(p)$ by:

$\mathbf{x_h} = (x_{h1}, \ldots, x_{hm_h}) \in EB_h(p)$ if and only if $\mathbf{x_h} \in B_h(p)$ and there is no $\mathbf{x'_h} \in B_h(p)$ such that

$$U_{hm}(\mathbf{x'_h}) \ge U_{hm}(\mathbf{x_h}) \text{ for all } m = 1, \ldots, m_h;$$
$$U_{hm}(\mathbf{x'_h}) > U_{hm}(\mathbf{x_h}) \text{ for some } m = 1, \ldots, m_h.$$

Thus efficient choice by the household refers to the individual consumption and welfare of its members, not merely to the aggregate consumption bundle of the household.

5.2.4 Feasibility and Optimality

An allocation $(\mathbf{x}, \mathbf{y}) = ((x_i)_{i \in I}, (y_j)_{j \in J}) \in \mathcal{X} \times \mathcal{Y}$ is **feasible** if

$$\sum_{i \in I} x_i = \omega + \sum_{j \in J} y_j. \tag{5.1}$$

A feasible allocation $(\mathbf{x}, \mathbf{y}) = ((x_i)_{i \in I}, (y_j)_{j \in J})$ is **Pareto optimal** if there is no other feasible allocation $(\mathbf{x}', \mathbf{y}') = ((x'_i)_{i \in I}, (y'_j)_{j \in J})$ such that

$$U_{hm}(\mathbf{x'_h}) \ge U_{hm}(\mathbf{x_h}) \text{ for all } h = 1, \ldots, n;\ m = 1, \ldots, m_h;$$
$$U_{hm}(\mathbf{x'_h}) > U_{hm}(\mathbf{x_h}) \text{ for some } h = 1, \ldots, n;\ m = 1, \ldots, m_h.$$

General Equilibrium: A **competitive equilibrium** is a triple $(p; (\mathbf{x}, \mathbf{y}))$ consisting of a price system p and an allocation $(\mathbf{x}, \mathbf{y}) = ((x_i)_{i \in I}, (y_j)_{j \in J})$ such that

1. $y_j \in Y_j(p)$ for all $j \in J$;
2. $\mathbf{x_h} \in EB_h(p)$ for all $h \in H$;
3. $(\mathbf{x}, \mathbf{y})$ is feasible, i.e., it satisfies (5.1).

In a general equilibrium, each firm maximizes profits, every household makes an efficient choice under its budget constraint and markets clear. We obtain a **first welfare theorem** for economies with multi-member households and production:

Proposition 5.1 *Let* $(p; (\mathbf{x}^*, \mathbf{y}^*))$ *be a competitive equilibrium such that*

$$p * \mathbf{x_h} = \mathrm{w}_h(p) \textit{ for all } h = 1, \ldots, n;\ \mathbf{x_h} \in EB_h(p). \tag{5.2}$$

Then $(\mathbf{x}^*, \mathbf{y}^*)$ *is a Pareto optimal allocation.*

Proof Suppose $(\mathbf{x}^*, \mathbf{y}^*)$ is not Pareto optimal. Then there exists a feasible allocation $(\mathbf{x}', \mathbf{y}')$ such that

$$\begin{aligned} U_{hm}(\mathbf{x}'_\mathbf{h}) &\geq U_{hm}(\mathbf{x}^*_\mathbf{h}) \ \text{ for all } h = 1, \ldots, n;\ m = 1, \ldots, m_h; \\ U_{hm}(\mathbf{x}'_\mathbf{h}) &> U_{hm}(\mathbf{x}^*_\mathbf{h}) \ \text{ for some } h = 1, \ldots, n;\ m = 1, \ldots, m_h. \end{aligned}$$

Hence there exists at least one household h with

$$\begin{aligned} U_{hm}(\mathbf{x}'_\mathbf{h}) &\geq U_{hm}(\mathbf{x}^*_\mathbf{h}) \ \text{ for all } m = 1, \ldots, m_h; \\ U_{hm}(\mathbf{x}'_\mathbf{h}) &> U_{hm}(\mathbf{x}^*_\mathbf{h}) \ \text{ for some } m = 1, \ldots, m_h. \end{aligned}$$

Since $\mathbf{x}^*_\mathbf{h} \in EB_h(p)$, we get $\mathbf{x}'_\mathbf{h} \notin B_h(p)$ for such an h:

$$p * \mathbf{x}'_\mathbf{h} > \mathrm{w}_h(p). \tag{5.3}$$

(5.3) holds for all such households. For the remaining households, $U_{hm}(\mathbf{x}'_\mathbf{h}) = U_{hm}(\mathbf{x}^*_\mathbf{h})$ for all $m = 1, \ldots, m_h$. If $\mathbf{x}'_\mathbf{h} \in B_h(p)$, then $\mathbf{x}'_\mathbf{h} \in EB_h(p)$ because of $\mathbf{x}^*_\mathbf{h} \in EB_h(p)$ and, consequently, $p * \mathbf{x}'_\mathbf{h} = \mathrm{w}_h(p)$ because of (5.2). If $\mathbf{x}'_\mathbf{h} \notin B_h(p)$, then (5.3) holds. In any case,

$$p * \mathbf{x}'_\mathbf{h} \geq \mathrm{w}_h(p). \tag{5.4}$$

(5.3) and (5.4) yield

$$p \sum_{i \in I} x'_i = \sum_{h \in H} p * \mathbf{x}'_\mathbf{h} > \sum_{h \in H} \mathrm{w}_h(p).$$

Now $py^*_j \geq py'_j$ for each $j \in J$. Hence

$$\begin{aligned} p \sum_{i \in I} x'_i &> \sum_{h \in H} \mathrm{w}_h(p) \\ &= \sum_{h \in H} \left[p\omega_h + \sum_{j \in J} \theta_{hj} p y^*_j \right] \\ &= p \sum_{h \in H} \omega_h + \sum_{h \in H} \sum_{j \in J} \theta_{hj} p y^*_j \end{aligned}$$

$$\geq p\omega + \sum_{h\in H}\sum_{j\in J}\theta_{hj}py'_j$$

$$= p\omega + p\sum_{j\in J}y'_j\sum_{h\in H}\theta_{hj}$$

$$= p\omega + p\sum_{j\in J}y'_j = p\left[\omega + \sum_{j\in J}y'_j\right],$$

contradicting $\sum_{i\in I}x'_i = \omega + \sum_{j\in J}y'_j$, the feasibility of $(\mathbf{x}', \mathbf{y}')$. Hence to the contrary, $(\mathbf{x}^*, \mathbf{y}^*)$ has to be Pareto optimal. □

5.3 Existence

In contrast to the existence proofs for pure exchange economies in Gersbach and Haller (1999) and Sato (2009), who take the excess demand approach, we rely on the simultaneous optimization or social equilibrium approach of Arrow and Debreu (1954). In order to make the latter applicable, we replace each household's consumption set $\mathcal{X}_h$ by the aggregate consumption set $\mathcal{A}_h = \mathbb{R}^\ell_+$ so that (aggregate) consumption bundles and production plans have equal dimension. We also need preferences on $\mathcal{A}_h = \mathbb{R}^\ell_+$ that reflect household preferences. We are going to define a utility function $V_h : \mathcal{A}_h \to \mathbb{R}$ with the desired properties.

5.3.1 Preferences on Aggregate Household Consumption

Consider a household h with members $i = hm, m = 1, \ldots, m_h$ and household consumption set $\mathcal{X}_h = \prod_{m=1}^{m_h} X_{hm}$. We introduce the notation $\mathcal{A}_h = \mathbb{R}^\ell_+$ for the household's aggregate consumption set. We further define a canonical mapping $A_h : \mathcal{X}_h \longrightarrow \mathcal{A}_h$ that assigns to each household consumption plan $\mathbf{x_h} = (x_{h1}, \ldots, x_{hm_h})$ the aggregate consumption $A_h(\mathbf{x_h}) = \sum_m x_{hm}$. For each $a_h \in \mathcal{A}_h$, we are interested in the inverse image $A_h^{-1}(a_h)$, the household consumption plans that give rise to the aggregate consumption a_h for household h. For all $a_h \in \mathcal{A}_h$, $A_h^{-1}(a_h) \neq \emptyset$, since $(a_h, 0, \ldots, 0) \in A_h^{-1}(a_h)$ and $(a_h/m_h, \ldots, a_h/m_h) \in A_h^{-1}(a_h)$, for example. Two properties obviously hold:

(P1) The correspondence $A_h^{-1} : \mathcal{A}_h \twoheadrightarrow \mathcal{X}_h$ is convex and compact valued.
(P2) The correspondence $A_h^{-1} : \mathcal{A}_h \twoheadrightarrow \mathcal{X}_h$ is continuous.

Next fix for household h a utilitarian social welfare function $W_h : \mathcal{X}_h \to \mathbb{R}$ of the form

$$W_h(\mathbf{x_h}) = \sum_{m=1}^{m_h} c_{hm} \cdot U_{hm}(\mathbf{x_h}) \ \text{ for all } \mathbf{x_h} \in \mathcal{X}_h$$

where $c_h = (c_{h1}, \ldots, c_{hm_h}) \in \mathbb{R}^{m_h}_{++}$. If each U_{hm} is continuous, then W_h is continuous and because of the compactness and non-emptiness of $A_h^{-1}(a_h)$,

$$V_h(a_h) = \max_{\mathbf{x_h} \in A_h^{-1}(a_h)} W_h(\mathbf{x_h}) \tag{5.5}$$

is well defined for all $a_h \in \mathcal{A}_h$. Moreover:

(P3) If each U_{hm} is continuous and concave, then $V_h : \mathcal{A}_h \to \mathbb{R}$ is continuous and concave.

Namely, W_h is continuous. In addition, (P1) and (P2) hold. Hence continuity of V_h follows from Berge's Maximum Theorem. If each U_{hm} is concave, then W_h is concave as well. Now let $a_h, a'_h \in \mathcal{A}_h$ and $\lambda \in (0, 1)$. There exist $\mathbf{x_h} \in A_h^{-1}(a_h)$ and $\mathbf{x'_h} \in A_h^{-1}(a'_h)$ such that $V_h(a_h) = W_h(\mathbf{x_h})$ and $V_h(a'_h) = W_h(\mathbf{x'_h})$. Further, $\lambda \cdot \mathbf{x_h} + (1-\lambda) \cdot \mathbf{x'_h} \in A_h^{-1}(\lambda \cdot a_h + (1-\lambda) \cdot a'_h)$ and $W_h(\lambda \cdot \mathbf{x_h} + (1-\lambda) \cdot \mathbf{x'_h}) \geq \lambda \cdot W_h(\mathbf{x_h}) + (1-\lambda) \cdot W_h(\mathbf{x'_h})$. Therefore, $V_h(\lambda \cdot a_h + (1-\lambda) \cdot a'_h) \geq W_h(\lambda \cdot \mathbf{x_h} + (1-\lambda) \cdot \mathbf{x'_h}) \geq \lambda \cdot W_h(\mathbf{x_h}) + (1-\lambda) \cdot W_h(\mathbf{x'_h}) = \lambda \cdot V_h(a_h) + (1-\lambda) \cdot V_h(a'_h)$. This shows concavity of V_h.

5.3.2 *Equilibrium Existence Result*

We are now prepared to state an equilibrium existence result. Let $Y = \sum_j Y_j$ denote the aggregate production set.

Proposition 5.2 *A competitive equilibrium exists if*
for every consumer $i = hm$,
(C) U_i is continuous and concave;
for every household h,
(H) $\omega_h \gg 0$ and local non-satiation holds;
for every firm j,
(F) $0 \in Y_j$;
for the aggregate production set Y,
(Y) Y is closed and convex; $Y \cap (-Y) = \{0\}$; $\mathbb{R}^\ell_- \subseteq Y$.

Proof Suppose (C) for all consumers i, (H) for all households h, (F) for all firms j, and (Y) for the aggregate productions set. Fix a utilitarian social welfare function $W_h : \mathcal{X}_h \to \mathbb{R}$ for every household h. Consider the finite Arrow-Debreu economy $\mathfrak{E} = ((\mathcal{A}_h, V_h, \omega_h)_{h\in H}, (\theta_{hj})_{(h,j)\in H\times J}, (Y_j)_{j\in J})$ where $\mathcal{A}_h = \mathbb{R}^\ell_+$ and V_h is given by (5.5) for $h \in H$. Then:

(i) Each $\mathcal{A}_h$ is closed, convex, and bounded from below.
(ii) Each "consumer" h is locally non-satiated.
Namely, let $a_h \in \mathcal{A}_h$ and $\epsilon > 0$. Let $\mathbf{x_h} \in A_h^{-1}(a_h)$ with $V_h(a_h) = W_h(\mathbf{x_h})$. Because of (H), local non-satiation holds for household h: There exists $\mathbf{x'_h} \in \mathcal{X}_h$

with $\| \mathbf{x_h} - \mathbf{x'_h} \|_{m_h\ell} < \epsilon/m_h$ and $(U_i(\mathbf{x'_h}))_{i\in h} > (U_i(\mathbf{x_h}))_{i\in h}$. Then $W_h(\mathbf{x'_h}) > W_h(\mathbf{x_h})$. Let $a'_h \equiv A_h(\mathbf{x'_h}) \in \mathcal{A}_h$. It follows $\| a_h - a'_h \|_\ell = \| A_h(\mathbf{x_h}) - A_h(\mathbf{x'_h}) \|_\ell = \| \sum_{m=1}^{m_h}(x_{hm} - x'_{hm}) \|_\ell \le \sum_{m=1}^{m_h} \| x_{hm} - x'_{hm} \|_\ell \le m_h \| \mathbf{x_h} - \mathbf{x'_h} \|_{m_h\ell} < \epsilon$. Moreover, $V_h(a'_h) \ge W_h(\mathbf{x'_h}) > W_h(\mathbf{x_h}) = V_h(a_h)$. Hence there exists $a'_h \in \mathcal{A}_h$ such that $\| a_h - a'_h \|_\ell < \epsilon$ and $V_h(a'_h) > V_h(a_h)$. This shows that "consumer" h is locally non-satiated.

Further:

(iii) Each V_h is continuous and concave, by (P3).
(iv) $\omega_h \gg 0$ for all h.
(v) $0 \in Y_j$ for all j.
(vi) Y is closed and convex; $Y \cap (-Y) = \{0\}$; $\mathbb{R}^\ell_- \subseteq Y$.

(i)–(vi) imply that $\mathfrak{E}$ satisfies the hypothesis of Theorem 5 of Debreu (1982). Therefore, the economy $\mathfrak{E}$ has a competitive equilibrium $((a^*_h)_{h\in H}, (y^*_j)_{j\in J}, p^*)$ in the sense of Debreu. For each $h \in H$, choose $\mathbf{x^*_h} = (x^*_{h1}, \ldots, x^*_{hm_h}) \in A_h^{-1}(a^*_h)$ with $V_h(a^*_h) = W_h(\mathbf{x^*_h})$. Let $\mathbf{x}^* = (x^*_i)_{i\in I}$ and $\mathbf{y}^* = (y^*_j)_{j\in J}$. We claim that $(p^*; (\mathbf{x}^*, \mathbf{y}^*))$ is a competitive equilibrium of the economy with multi-member households $h \in H$.

1. $y^*_j \in Y_j(p^*)$ for all $j \in J$, by the definition of a competitive equilibrium of $\mathfrak{E}$.
2. $\mathbf{x^*_h} \in EB_h(p^*)$ for all $h \in H$. Namely, $V_h(a^*_h) = \max\{V_h(a_h) : a_h \in \mathcal{A}_h,\ p^*a_h \le \mathrm{w}_h(p^*)\}$. Since $a^*_h = A_h(\mathbf{x^*_h})$, we get $p^* * \mathbf{x^*_h} = p^*a^*_h \le \mathrm{w}_h(p^*)$ and, therefore, $\mathbf{x^*_h} \in B_h(p^*)$. If $\mathbf{x^*_h} \notin EB_h(p^*)$, then there exists $\mathbf{x'_h} \in B_h(p^*)$ such that $U_{hm}(\mathbf{x'_h}) \ge U_{hm}(\mathbf{x^*_h})$ for all $m = 1, \ldots, m_h$ and $U_{hm}(\mathbf{x'_h}) > U_{hm}(\mathbf{x^*_h})$ for some $m = 1, \ldots, m_h$. Choose such an $\mathbf{x'_h}$. Then $W_h(\mathbf{x'_h}) > W_h(\mathbf{x^*_h})$. Hence for $a'_h = A_h(\mathbf{x'_h})$: $a'_h \in \mathcal{A}_h$, $p^*a'_h = p^* * \mathbf{x'_h} \le \mathrm{w}_h(p^*)$, $V_h(a'_h) \ge W_h(\mathbf{x'_h}) > W_h(\mathbf{x^*_h}) = V_h(a^*_h)$, contradicting the fact that $V_h(a^*_h) = \max\{V_h(a_h) : a_h \in \mathcal{A}_h,\ p^*a_h \le \mathrm{w}_h(p^*)\}$. Therefore, $\mathbf{x^*_h} \in EB_h(p^*)$ has to hold.
3. $(\mathbf{x}^*, \mathbf{y}^*)$ is feasible, i.e., it satisfies (5.1). Namely, $\sum_{h\in H} a^*_h = \omega + \sum_{j\in J} y^*_j$ implies $\sum_{i\in I} x^*_i = \sum_{h\in H}\sum_{m=1}^{m_h} x^*_{hm} = \sum_{h\in H} A_h(\mathbf{x^*_h}) = \sum_{h\in H} a^*_h = \omega + \sum_{j\in J} y^*_j$ and, thus, (5.1).

We have shown that $(p^*; (\mathbf{x}^*, \mathbf{y}^*))$ satisfies conditions 1.-3. of a competitive equilibrium for the economy with multi-member households $h \in H$. This demonstrates the claim and completes the proof. □

5.4 Ramifications

General equilibrium analysis often includes a second welfare theorem, core inclusion, core equivalence, and related issues. We are going to briefly discuss these topics as well as household production.

5.4.1 Second Welfare Theorem

Proposition 6 of Gersbach and Haller (2001) asserts validity of a second welfare theorem for a pure exchange economy with fixed household structure. The proof of the proposition applies the separating hyperplane theorem. Like similar proofs in the literature, it can easily incorporate production.

5.4.2 Core Theory

Haller (2000) presents a H-core inclusion result where in the definition of the H-core or household core only unions of households in H qualify as coalitions. Again, the budget exhaustion property (5.2) proves instrumental. There is a sizeable literature on coalition-production economies, e.g. Böhm (1974) and Hildenbrand (1974, Chap. 4), where each coalition is endowed with its own technology. This approach provides an elegant way to extend the methods developed for pure exchange economies, but tends to ignore individual ownership of means of production. In other cases, individual private property in firms does not fully apply. Debreu and Scarf (1963) assume that production technologies are publicly available and exhibit constant returns to scale so that issues related to corporate control are absent. Allingham (1975) and Aliprantis et al. (1987) assume divisibility of technologies and that each shareholder controls a fraction of the firm's technology which avoids conflicts among shareholders.

In general economies with production and private property, the question arises when and how a coalition of consumers or households can alter the production plan of a firm that is not entirely owned by the coalition. First attempts to deal with this intricate question and to take fully into account the ramifications of private property have been made by Haller (1991) and Xiong and Zheng (2007). This complex issue is left to future research.

5.4.3 Household Production

By most accounts, household production creates substantial value in most economies. It was the article of Becker (1965) that positioned the use of time and household production firmly within economic theory. At the micro level, Becker's work and a rich subsequent literature (see e.g. Apps and Rees (2009) for a discussion) demonstrate that household production constitutes a major determinant of household welfare. For those reasons, a comprehensive account and description of household activities and intra-household allocation ought to include household production.[4]

[4]At the macro level, the value of household production is significant and could be around 35% of GDP in developed countries (see Apps and Rees (2009), p. 32).

Delineating household production in general equilibrium frameworks, however, proves difficult. The most stringent definition would require that the household uses its own factors of production to produce goods for its own consumption only. But households which are autarkic with respect to all factors of production barely exist. For example, to bake a cake, most of the basic ingredients are typically purchased in the market. A less stringent definition requires that labor and capital are owned by the household whereas intermediate products can be obtained in the market. Yet even then, households living in rental housing, for example, would not qualify for household production.

Within a general equilibrium framework, Gilles and Diamantaras (2003) assume that each consumer is endowed with his own home production set. They distinguish between tradeable and non-tradeable commodities. A consumer's productive activity is only considered household or home production if the output consists of non-tradeable commodities. Individuals can own and consume non-tradeable commodities, but they are restricted to consuming their home produced quantities of these non-tradeables. Now almost any commodity is tradeable at some time in some place. But that is exactly Gilles's and Diamantaras's point: tradeability is an endogenous, temporal and local feature. Gilles and Diamantaras (2003) demonstrate that welfare analysis involving transfers—adopting a valuation equilibrium concept to be precise—can be performed with their formalization of home production and its connection with tradeability. Yet showing existence of a competitive equilibrium that *ceteris paribus* respects private property rights seems beyond reach.

5.5 Final Remarks

Proposition 5.1 means that, by and large, competitive exchange among profit maximizing firms and multi-member households satisfying the collective rationality model yields Pareto optimal allocations. Obviously, local non-satiation of households prevails and a fortiori the budget exhaustion property (5.2) holds if all individuals exhibit strict monotonicity in own consumption and all intra-household consumption externalities are nonnegative. Example 3.1 in Haller (2000) illustrates that local non-satiation of households can still hold if all individuals in multi-member households experience specific negative consumption externalities. But Example 3.3 of Haller (2000), reproduced in Chap. 4, demonstrates that certain negative consumption externalities can lead to violation of (5.2) (and of local non-satiation) and yield suboptimal equilibrium allocations. Still, equilibrium allocations are always weakly Pareto optimal without any further assumptions.

We have noted that extension of the second welfare theorem to production also proves rather straightforward. Overall, we can conclude that production can easily be integrated into the welfare analysis we have conducted so far for pure exchange economies with multi-member households.

Proposition 5.2 states existence of a competitive equilibrium for a finite economy with profit maximizing firms and multi-member households satisfying the collective rationality model, under almost standard assumptions. The only exception is the assumption of local non-satiation of households which possibly can be replaced by weaker but less transparent assumptions. Without any assumption of this kind, one can expect an equilibrium with free disposal at best.

There are two more instances where we introduce a production sector. First, a closed model of the economy with an active labor market requires a production sector—if one disregards labor employed by households. In Gersbach and Haller (2012), we introduce a production sector to endogenize labor demand and wage rates. Second, while the inclusion of production tends to be straightforward, an active production sector can alter welfare conclusions. In Chap. 13, we show by example that the presence of a linear technology can lead to inefficient equilibrium outcomes when households are prone to make inefficient net trades whereas *ceteris paribus* efficient equilibrium allocations obtain in the case of pure exchange. It is well known that introduction of a constant returns to scale technology can have other qualitative consequences as well, for instance cause indeterminacy of equilibrium while gross substitutes prevail in the consumption sector (see Kehoe (1985)).

References

Aliprantis, C., Brown, D., & Burkinshaw, D. (1987). Edgeworth equilibria in production economies. *Journal of Economic Theory, 43*, 252–291.

Allingham, M. (1975). *General equilibrium*. London and Basingstoke: The MacMillan Press.

Apps, P. F., & Rees, R. (2009). *Public economics and the household*. Cambridge, UK: Cambridge University Press.

Arrow, K. J., & Debreu, G. (1954). Existence of an equiibrium for a competitive economy. *Econometrica, 22*, 265–290.

Becker, G. S. (1965). A theory of the allocation of time. *Economic Journal, 75*, 493–517.

Böhm, V. (1974). The core of an economy with production. *Review of Economic Studies, 41*, 429–436.

Chiappori, P.-A. (1988). Rational household labor supply. *Econometrica, 56*, 63–89.

Chiappori, P.-A. (1992). Collective labor supply and welfare. *Journal of Political Economy, 100*, 437–467.

Debreu, G. (1982). Existence of competitive equilibrium. In K. J. Arrow & M. D. Intriligator (Eds.), *Handbook of mathematical economics* (Vol. 2). Amsterdam: North-Holland.

Debreu, G., & Scarf, H. (1963). A limit theorem on the core of an economy. *International Economic Review, 4*, 235–246.

Gersbach, H., & Haller, H. (1999). Allocation among multi-member households: Issues, cores and equilibria. In A. Alkan, C. D. Aliprantis, & N. C. Yannelis (Eds.), *Current trends in economics: Theory and applications*. Heidelberg: Springer-Verlag, Berlin.

Gersbach, H., & Haller, H. (2001). Collective decisions and competitive markets. *Review of Economic Studies, 68*, 347–368.

Gersbach, H., & Haller, H. (2005). When inefficiency begets efficiency. *Economic Theory, 25*, 105–121.

Gersbach, H., & Haller, H. (2009). Bargaining power and equilibrium consumption. *Social Choice and Welfare, 33*, 665–690.

Gersbach, H., & Haller, H. (2010). Club theory and household formation. *Journal of Mathematical Economics, 46*, 715–724.

Gersbach, H., & Haller, H. (2011). Bargaining cum voice. *Social Choice and Welfare*, *36*, 199–225.

Gersbach, H., & Haller, H. (2012). "Hard Workers" and labor restrictions. *Journal of Population Economics*, *25*, 469–494.

Gilles, R. P., & Diamantaras, D. (2003). To trade or not to trade: Economies with a variable number of tradeables. *International Economic Review*, *44*, 1173–1204.

Gori, M. (2010). Endogenous household formation and inefficiency in a general equilibrium model. Working Paper, Dipartimento di Matematica per le Decisioni, Università degli Studi di Firenze.

Gori, M., & Villanacci, A. (2011). A bargaining model in general equilibrium. *Economic Theory*, *46*, 327–375.

Haller, H. (1991). Corporate production and shareholder cooperation under uncertainty. *International Economic Review*, *32*, 823–842.

Haller, H. (2000). Household decisions and equilibrium efficiency. *International Economic Review*, *41*, 835–847.

Hildenbrand, W. (1974). *Core and equilibria of a large economy*. Princeton, NJ: Princeton University Press.

Kehoe, T. J. (1985). Multiplicity of equilibria and comparative statics. *Quarterly Journal of Economics*, *100*, 119–147.

Lundberg, S., & Pollak, R. A. (1993). Separate spheres bargaining and the marriage market. *Journal of Political Economy*, *101*, 988–1011.

Lundberg, S., & Pollak, R. A. (1994). Non-cooperative bargaining models of marriage. *American Economic Review, Papers and Proceedings*, *84*, 132–137.

Sato, N. (2009). Existence of competitive equilibrium in economies with multi- member households. *Economics Bulletin*, *29*, 1760–1771.

Xiong, S., & Zheng, C. Z. (2007). Core equivalence theorem with production. *Journal of Economic Theory*, *137*, 246–270.

Chapter 6
General Equilibrium with Variable Household Structure

In the present chapter, we resume our inquiry into the allocation of commodities and consumers through the interaction of two allocation mechanisms for households: (b) collective decisions and (c) competitive commodity markets. Boundaries between collective decisions and markets are determined by the household structure. The principal innovation of this chapter are variable household structures that allow flexible boundaries between collective decisions and markets for the allocation of commodities. We are mainly interested in the efficiency properties and in the decentralization possibilities of the dual allocation mechanism.

6.1 Consumers and Households; Commodities and Allocations

In order to simplify the formal presentation and to facilitate the exposition of several examples, we first introduce a general and convenient notation for the comparison of vectors.

Consider any $L \in \mathbb{N}$, $a = (a_1, \ldots, a_L)$, $b = (b_1, \ldots, b_L) \in \mathbb{R}^L$, and $f : \mathbb{R}^L \to \mathbb{R}$. By $a \geq b$, we mean $a_l \geq b_l$ for all $l = 1, \ldots, L$. By $a \gg b$ we mean $a_l > b_l$ for all l. Finally, $a > b$ stands for $a \geq b$, $a \neq b$. The function f is called non-decreasing, if for any $a, b \in \mathbb{R}^L$, $a \geq b$ implies $f(a) \geq f(b)$. It is increasing, if for any $a, b \in \mathbb{R}^L$, $a \gg b$ implies $f(a) > f(b)$. It is strictly increasing, if for any $a, b \in \mathbb{R}^L$, $a > b$ implies $f(a) > f(b)$. If F is a finite set and $\mathbf{x_F} = (x_j)_{j \in F} \in (\mathbb{R}^L)^F$, then the aggregate of the vector $\mathbf{x_F}$ over the "population" F is denoted $\bar{\mathbf{x}}_F$ and defined as $\bar{\mathbf{x}}_F = \sum_{j \in F} x_j \in \mathbb{R}^L$.

H. Gersbach and H. Haller, *Groups and Markets*,
DOI 10.1007/978-3-319-60516-6_6

The primitive concepts are defined and assumed as in Chap. 4. We deal with a finite set of **consumers** or **individuals**, represented by a set $I = \{1, \ldots, N\}$. A generic consumer is denoted i or j. The population I is partitioned into households, i.e., there exists a partition P of I into non-empty subsets. We call any such partition P a **household structure** in I. Let $\mathcal{P}$ denote the set of all household structures in I.

There exists a finite number $\ell \geq 1$ of commodities. Thus the commodity space is $\mathbb{R}^\ell$. Each commodity is a private good. Each consumer $i \in I$ has consumption set $X_i = \mathbb{R}^\ell_+$, so that the **commodity allocation space** is $\mathcal{X} \equiv \prod_{j \in I} X_j$. Let $\mathbf{x} = (x_i)$, $\mathbf{y} = (y_i)$ denote generic elements of $\mathcal{X}$.

An **allocation** is a pair $(\mathbf{x}; P) \in \mathcal{X} \times \mathcal{P}$ specifying the consumption bundle and household membership of each consumer.

For a potential household $h \subseteq I, h \neq \emptyset$, set $\mathcal{X}_h = \prod_{i \in h} X_i$, the consumption set for household h. $\mathcal{X}_h$ has generic elements $\mathbf{x_h} = (x_i)_{i \in h}$. If $\mathbf{x} \in \mathcal{X}$ is a commodity allocation, then consumption for household h is the restriction of $\mathbf{x} = (x_i)_{i \in I}$ to h, $\mathbf{x_h} = (x_i)_{i \in h}$. If $(\mathbf{x}; P)$ is an allocation, then a household $h \in P$ attains the household consumption $\mathbf{x_h} \in \mathcal{X}_h$.

The economic units endowed with resources are households rather than individuals. Notice, however, that in an environment with endogenous household formation, each singleton $\{i\}$ is a potential one-person household with its own endowment. For a potential household $h \subseteq I, h \neq \emptyset$, its **endowment** is a commodity bundle $\omega_h \in \mathbb{R}^\ell$, $\omega_h \geq 0$. A special case is

(IPR) Individual Property Rights: $\omega_h = \sum_{i \in h} \omega_{\{i\}}$ for each household h.

In general, the social endowment with resources depends on the household structure. Namely, if the household structure $P \in \mathcal{P}$ is in place, then the **social endowment** is

$$\omega_P \equiv \sum_{h \in P} \omega_h.$$

A different household structure can yield a different social endowment. Allowing the endowment of a household to differ from the sum of endowments of the potential one-person households formed by its members can be interpreted as resource costs of setting up households or, in the opposite direction, as economies of scale enjoyed by larger households.

We call an allocation $(\mathbf{x}; P) \in \mathcal{X} \times \mathcal{P}$ **feasible**, if

$$\sum_{i \in I} x_i = \omega_P. \tag{6.1}$$

$\mathcal{F}$ denotes the set of feasible allocations. Notice that if $(\mathbf{x}; P) \in \mathcal{F}$, then $\mathbf{x}$ is called a P-**feasible commodity allocation**. $\mathcal{X}(P)$ denotes the set of P-feasible commodity allocations.

With *free disposal* of resources, the feasibility requirement could and should be relaxed. We call an allocation $(\mathbf{x}; P) \in \mathcal{X} \times \mathcal{P}$ **weakly feasible**, if

$$\sum_{i \in I} x_i \leq \omega_P. \tag{6.2}$$

6.2 Preferences and Welfare

We make the **Assumption of Household-Specific Preferences (HSP)** of Chap. 3 with individual utility representations $U_i : \mathcal{A}_i \to \mathbb{R}, i \in I$, where $\mathcal{X}^* = \bigcup_{h \in \mathcal{H}} \mathcal{X}_h$ and $\mathcal{A}_i = \{(\mathbf{x_h}; h) \in \mathcal{X}^* \times \mathcal{H} : h \in \mathcal{H}_i, \mathbf{x_h} \in \mathcal{X}_h\}$ for $i \in I$. With a fixed, exogenously given household structure P, household membership can be considered part of an individual's identity in which case household affiliation may be dropped as argument of the individual's utility function. Thus, one obtains the utility representation $u_i : \mathcal{X}_h \to \mathbb{R}$, given by $u_i(\mathbf{x_h}) = U_i(\mathbf{x_h}; h)$ for $\mathbf{x_h} \in \mathcal{X}_h$ and $i \in h \in P$.

We shall resort to some of the special properties of preferences distinguished in Chap. 3: Pure Group Externalities (PGE), Group-Size Externalities (GSE), Monotonicity (MON), Strict Monotonicity (SMO), Non-Negative Externalities (NNE), Non-Positive Externalities (NPE), Separable Externalities (SEP), Local Anonymity (LAN), and Global Anonymity (GAN).

Household preferences are represented by the preference relation $\succsim_h$ on $\mathcal{X}_h$, given by

$$\mathbf{x_h} \succsim_h \mathbf{y_h} \Longleftrightarrow [U_i(\mathbf{x_h}; h) \geq U_i(\mathbf{y_h}; h) \, \forall i \in h]$$

for $\mathbf{x_h}, \mathbf{y_h} \in \mathcal{X}_h$. Occasionally, it proves useful to employ the strict preference relation $\gg_h$ on $\mathcal{X}_h$ given by

$$\mathbf{x_h} \gg_h \mathbf{y_h} \Longleftrightarrow [U_i(\mathbf{x_h}; h) > U_i(\mathbf{y_h}; h) \, \forall i \in h]$$

for $\mathbf{x_h}, \mathbf{y_h} \in \mathcal{X}_h$.

We will further work with the following two properties from Gersbach and Haller (2001) that apply to households rather than individual members, the **redistribution property** and **local non-satiation** to multi-person households:

Definition (*Redistribution Property (RP)*) The Redistribution Property holds for household h if for any two bundles $\mathbf{x_h}, \mathbf{y_h} \in \mathcal{X}_h$ with $\mathbf{y_h} \succ_h \mathbf{x_h}$, there exists $\mathbf{z_h} \in \mathcal{X}_h$ such that $\sum_{i \in h} z_i = \sum_{i \in h} y_i$ and $\mathbf{z_h} \gg_h \mathbf{x_h}$.

Notice that Social Monotonicity (SM) à la Dufwenberg et al. (2011) is reminiscent of the Redistribution Property. (SM) differs from (RP) in that it applies to the entire population without a specific household structure; preferences exhibit well-being externalities, that is, there exist functions $m_i : \mathbb{R}^{\ell}_{+} \to \mathbb{R}$, $i \in I$, and $V_i : \mathbb{R}^n \to \mathbb{R}$, $i \in I$, with V_i strictly increasing in the i-th component, such that $U_i(\mathbf{x}) = V_i(m_1(x_1), \ldots, m_n(x_n))$ for $\mathbf{x} = (x_1, \ldots, x_n) \in \mathcal{X}$; $\mathbf{z} > \mathbf{x}$ (in our notation).

At first sight, the Redistribution Property appears to be a weak constraint on the preferences of household members. Indeed, (RP) is fulfilled in many circumstances. However, (RP) can be violated when consumption externalities are strong and positive. In such circumstances, redistributing commodities from one household member to others may not improve the utility of the individuals receiving those consumption goods. Let us look at two examples where (RP) fails.

Example 6.1 Let $\ell = 1, h = \{1, 2\}$ and consumer preferences represented by the utility functions

$U_1(x_1, x_2; h) = (1 - x_1)^2$ and

$U_2(x_1, x_2; h) = x_2$.

Now consider $\mathbf{x} = (x_1, x_2) = (1, 1)$ and $\mathbf{y} = (y_1, y_2) = (1, 2)$.

Then $U_1(\mathbf{y}; h) = U_1(\mathbf{x}; h)$ and $U_2(\mathbf{y}; h) > U_2(\mathbf{x}; h)$.

But there is no way (no transfer) that would raise the welfare of consumer 1 above the status quo given by $\mathbf{x}$. □□

Obviously, this example exploits the fact that one of the consumers has a satiation consumption whereas the other consumer has none. But there are other violations of (RP). The next example relies on positive externalities such that the positive direct effect of a transfer is offset by the negative indirect impact via the externality.

Example 6.2 The model is identical with the one of Example 6.1, except for the preferences of consumer 1 which are now represented by

$U_1(x_1, x_2; h) = x_1 + x_2$.

Next consider $\mathbf{x} = (x_1, x_2) = (2, 2)$ and $\mathbf{y} = (y_1, y_2) = (1, 3)$.

Then $U_1(\mathbf{y}; h) = U_1(\mathbf{x}; h)$ and $U_2(\mathbf{y}; h) > U_2(\mathbf{x}; h)$.

Further for any $\mathbf{z} = (z_1, z_2) \in \mathcal{X}_h = \mathbb{R}^2_+$ with $z_1 + z_2 = y_1 + y_2$, $U_1(\mathbf{z}; h) = z_1 + z_2 = y_1 + y_2 = x_1 + x_2 = U_1(\mathbf{x}; h)$, i.e., consumer 1's welfare cannot be enhanced by a transfer from 2 to 1 or vice versa. □□

A sufficient condition for (RP) ought to rule out Examples 6.1 and 6.2. The following condition does achieve just this:

All utility functions are C^1 functions and exhibit strict monotonicity and not too positive externalities.

Here is a possible formal expression of *not too positive* externalities:

If $h \in \mathcal{H}$, $|h| \geq 2, i \in h$, then for all

$j \in h \backslash i, \mathbf{x_h} = (x_\iota)_{\iota \in h}, \mathbf{y_h} = (y_\iota)_{\iota \in h} \in \mathcal{X}_h$:

$\partial U_i(\mathbf{x_h}; h)/\partial x_i > (|h| - 1) \cdot \partial U_i(\mathbf{y_h}; h)/\partial y_j$.

This allows for (NPE), absence of externalities, or a moderate form of (NNE).

In the following definition, let $\| \cdot \|_{|h|\ell}$ denote the Euclidean norm in $\mathbb{R}^{|h|\ell}$.

Definition A household h is **locally non-satiated** if for every $\mathbf{x_h} \in \mathcal{X}_h$ and every $\epsilon > 0$, there exists $\mathbf{y_h} \in \mathcal{X}_h$ with $\| \mathbf{x_h} - \mathbf{y_h} \|_{|h|\ell} < \epsilon$ and $\mathbf{y_h} \succ_h \mathbf{x_h}$ where $\| \cdot \|_d$ denotes the Euclidean norm on $\mathbb{R}^d$.

We note that (RP) and local non-satiation are independent properties. A household may be locally non-satiated while (RP) fails, as illustrated by Example 6.2. Conversely, (RP) may hold, but the household may be locally satiated at some consumption bundle $\mathbf{x_h}$. To see this, consider $\ell = 1, h = \{1, 2\}, U_1((x_1, x_2); h) = U_2((x_1, x_2); h) = -(x_1 - x_2)^2$.

The conjunction of (MON) with absence of externalities is certainly sufficient for local non-satiation. It also suffices that for each household h, there exist a member $i(h) \in h$ and a commodity $c(h) \in \{1, \dots, \ell\}$ such that

- the welfare of $i(h)$ is increasing in $i(h)$'s consumption of commodity $c(h)$ and
- the welfare of all other household members is unaffected by $i(h)$'s consumption of commodity $c(h)$.

In particular, the hypothesis of the following variant of the second welfare theorem, stipulating (RP) plus local non-satiation, is satisfied in many instances. Incidentally, local non-satiation is not only instrumental to our formulation of the second welfare theorem. It also implies the budget exhaustion property introduced in the previous chapter. The proof is straightforward. On the other hand, budget exhaustion by itself does not imply local non-satiation as the next example demonstrates.
Given a household structure P, we can express the budget exhaustion property as follows:

(BE) Budget Exhaustion: For each household $h \in P$, any household consumption profile $\mathbf{x_h} \in \mathcal{X}_h$, and any price system $p \in \mathbb{R}^\ell$,

$$\boxed{\mathbf{x_h} \in EB_h(p) \Rightarrow p * \mathbf{x_h} = p \cdot \omega_h}$$

Example 6.3 Consider a single-person household h occupied by individual i. Let $\ell = 2$ and let $x_i = (x_i^1, x_i^2)$ denote a generic consumption bundle of consumer i whose preferences are represented by the continuous utility function $U_i(x_i^1, x_i^2) = f(x_i^1) + f(x_i^2)$ where

$$f(t) = \begin{cases} t^2 & \text{for } t > 4/3; \\ 4t - 2t^2 = 4t - 3t^2 + t^2 & \text{for } 0 \leq t \leq 4/3. \end{cases}$$

Then U_i is strictly concave on $[0, 4/3]^2$ with a unique local maximum at $x_i^* = (1, 1)$ and value 4. Also, U_i is strictly increasing, strictly convex and unbounded outside of $[0, 4/3]^2$. Hence (BE) holds for any budget set not containing x_i^* or having x_i^* as a boundary point. On the other hand, if x_i^* is an interior point of a budget set, then two possibilities arise. The first possibility is that the budget set is unbounded in which case the efficient budget set is empty and (BE) is trivially satisfied. A second possibility is that the corresponding price vector is strictly positive and the budget line intersects the diagonal at a point (t, t) with $t > 1$ and the axes at points $(a, 0)$ and $(0, b)$ with $\max\{a, b\} \geq 2t$. Hence under the budget constraint, utility is maximized at one or both of the two boundary points of the budget line, with value

$(\max\{a, b\})^2 \geq 4t^2 > 4$. Once more, (BE) holds true. Hence we have shown the coexistence of (BE) and local satiation. □□

In order to construct the example, we had to choose U_i convex on part of its domain. It is straightforward to see that (BE) (for arbitrary budget sets obtained by variation of prices and endowments) combined with concavity of the utility functions implies local non-satiation. To see this, consider a household h and $\mathbf{x_h} \in \mathcal{X}_h$. Choose $p \gg 0$ and ω_h so that $p * \mathbf{x_h} < p \cdot \omega_h$. Because of (BE), $\mathbf{x_h}$ does not belong to the corresponding efficient budget set. Thus there exists $\mathbf{y_h} \in \mathcal{X}_h$ with $p * \mathbf{y_h} \leq p \cdot \omega_h$ and

$$(U_i(\mathbf{y_h}; h))_{i \in h} > (U_i(\mathbf{x_h}; h))_{i \in h}.$$

Because of the concavity of the utility functions, such a $\mathbf{y_h}$ can be found arbitrarily close to $\mathbf{x_h}$.

The performance standard, that is which allocations qualify as "optimal" or "efficient", should depend on how much freedom a social planner is granted to allocate resources and people. Accordingly, we introduce two notions of Pareto optimality: constrained Pareto optimality and full Pareto optimality. The attribute "constrained" refers to a fixed household structure whereas "full" refers to a variable household structure.

Constrained Pareto Optimality. For any particular household structure P, we defined in Chap. 4 the concept of a contrained Pareto optimal commodity allocation with respect to P or P-optimal allocation. We maintain the notation $\mathcal{X}(P)$ for the set of P-feasible allocations and $\mathcal{M}(P)$ for the set of P-optimal allocations.

Full Pareto Optimality. The presumption here is that a social planner can allocate both commodities and consumers. An allocation $(\mathbf{x}; P)$ is called **(fully) Pareto optimal** or an **optimum optimorum**, if

(i) $(\mathbf{x}; P)$ is feasible and
(ii) there is NO feasible allocation $(\mathbf{x}'; P')$ satisfying

$$(U_i(\mathbf{x}'; P'))_{i \in I} > (U_i(\mathbf{x}; P))_{i \in I}. \tag{6.3}$$

Denote by $\mathcal{M}^*$ the set of Pareto optimal allocations. Write $(\mathbf{x}'; P') \succ (\mathbf{x}; P)$ as shorthand for (6.3) for any pair of feasible allocations $(\mathbf{x}'; P')$ and $(\mathbf{x}; P)$. Continuity of preferences implies existence of fully optimal allocations:

Proposition 6.1 (Gersbach and Haller (2001), Proposition 2) *Suppose all utility functions $U_i(\cdot; h)$, $i \in I$, $h \in \mathcal{H}_i$ are continuous. Then $\mathcal{M}^* \neq \emptyset$.*

The concept of full Pareto optimality allows the notion of an "optimal" household structure to be formalized. Namely set

$$\mathcal{P}^* \equiv \{P \in \mathcal{P} | \exists \mathbf{x} \in \mathcal{X} : (\mathbf{x}; P) \in \mathcal{M}^*\}.$$

Then a household structure P will be called **optimal**, if $P \in \mathcal{P}^*$, that is if it is part of an optimum optimorum (full Pareto optimum). Obviously, $(\mathbf{x}; P) \in \mathcal{M}^*$ implies $\mathbf{x} \in \mathcal{M}(P)$ and $P \in \mathcal{P}^*$.

6.3 Equilibrium Welfare

We adhere to the concept of P-equilibrium of Chap. 4, with the crucial difference that now the household structure is no longer treated as fixed. Given P, we found that with the budget exhaustion property, $\mathbf{x} \in \mathcal{M}(P)$ for any P-equilibrium $(p; \mathbf{x})$. But what can be said about the welfare properties of the entire allocation $(\mathbf{x}; P)$? By definition, $(\mathbf{x}; P) \in \mathcal{M}^*$ requires $P \in \mathcal{P}^*$. So suppose that $P \in \mathcal{P}^*$ and $\mathbf{x}$ is a P-equilibrium allocation. Can one then conclude that $(\mathbf{x}; P) \in \mathcal{M}^*$? In general, the answer is in the negative as Example 6.4 will demonstrate.

The current framework with a variable household structure makes it possible to address, in principle, the more general question which households should form. It also allows to address related welfare policy issues like who should be the designated recipients of transfers, households or individuals. The positive question which households will form has to wait until the next chapter when the household structure will be endogenized.

Like in Gersbach and Haller (2001), we take here a social planner approach with regard to the allocation of people to households. We allow a social planner to rearrange households and, thus, to choose an arbitrary household structure whereas the resource allocation is left to the market. Obviously, the existence of a social planner, powerful enough to impose detailed sociological choices on the entire population, is purely hypothetical and serves only as a benchmark.

One simple question one might address is whether two initially single consumers, say 1 and 2 would benefit from forming the two-person household $h = \{1, 2\}$. The answer depends on the status quo and to what extent the *ceteris paribus* clause applies. Suppose the status quo is the household structure $P^0 = \{\{i\} : i \in I\}$ consisting of all single households and the P^0-equilibrium $(p; \mathbf{x})$ which is unique up to price normalization. Suppose that *ceteris paribus* the household h forms and chooses $\mathbf{y_h} \in EB_h(p)$. If $U_1(\mathbf{y_h}; h) > U_1(x_1; \{1\})$ and $U_2(\mathbf{y_h}; h) > U_2(x_2; \{2\})$, then indeed, the household h might be likely to be formed—provided that in fact the equilibrium price system persists. The "state" $(p, \mathbf{x}; P^0)$ would violate one of the stability requirements we are going to put forward in the next chapter. One could envisage instead that a P^{00}-equilibrium $(p'; \mathbf{x}')$ different from $(p; \mathbf{x})$ obtains after the household structure has changed from P^0 to $P^{00} = P^0 \setminus \{\{1\}, \{2\}\} \cup \{\{1, 2\}\}$. Consumers 1 and 2 might not expect that equilibrium response. Still, one could pose the normative question whether they would benefit from household formation under those circumstances. Mohemkar-Kheirandish (2008) examines the question for several instructive examples, comparing allocations $(\mathbf{x}; P^0)$ and $(\mathbf{x}'; P^{00})$ where $\mathbf{x}$ a P^0-equilibrium allocation and $\mathbf{x}'$ is a P^{00}-equilibrium allocation.

Next let us return to our earlier concern, the welfare properties of an allocation of the form $(\mathbf{x}; P)$ where $P \in \mathcal{P}^*$ and $\mathbf{x}$ is a P-equilibrium allocation. Suppose (BE) is satisfied for any household structure $P \in \mathcal{P}$. Is it then the case that for any optimal household structure $P^* \in \mathcal{P}^*$, every P^*-equilibrium yields a fully Pareto optimal allocation of commodities and consumers? It turns out that the answer to this question is negative: Relying on the market for the allocation of commodities need not always result in an optimum optimorum, even under the favorable conditions postulated here. What if we settle for less? Is there some optimal household structure $P^{**} \in \mathcal{P}^*$ so that every P^{**}-equilibrium yields a fully Pareto optimal allocation of commodities and consumers? The answer is again negative. Settling for even less, could it be that for each optimal household structure $P^* \in \mathcal{P}^*$, there exists at least one P^*-equilibrium which yields a fully optimal allocation of commodities and consumers? Once again, the answer is negative. These negative findings all emanate from the following example, a shortened version of Example 2 in Gersbach and Haller (2001).

Example 6.4 Let $\ell = 1$ and $I = \{1, 2\}$. Consumer 1 experiences no externalities at all. Consumer 2 is subject to consumption externalities only.
Specifically, we assume
$U_1(\mathbf{x_h}; h) = x_1$ for $h = \{1\}, \{1, 2\}$;
$U_2(\mathbf{x_h}; h) = x_2$, in case $h = \{2\}$;
$U_2(\mathbf{x_h}; h) = x_2 + v(x_1)$, in case $h = \{1, 2\}$;
where $v(t) = 5t - 2t^2$ for $t \in \mathbb{R}$.

We observe that (BE) is satisfied no matter what the household structure is. We further assume that $\omega_1 = 2$, $\omega_2 = 1$, and $\omega_{\{1,2\}} = \omega_1 + \omega_2 = 3$. Denote
$P^0 = \{\{1\}, \{2\}\}$,
$P^1 = \{\{1, 2\}\}$.

Properties of P^0:

- The only P^0-equilibrium allocation is $\mathbf{x^0} = (2, 1)$. $(\mathbf{x^0}; P^0)$ is dominated by $(\mathbf{z^1}; P^1)$ where $\mathbf{z^1} = (2.1, 0.9)$.
- $P^0 \in \mathcal{P}^*$, since $(\mathbf{y^0}; P^0) \in \mathcal{M}^*$ where $\mathbf{y^0} = (3, 0)$.

Properties of P^1:

- $\mathbf{x^1} = (2.6, 0.4)$ is a P^1-equilibrium allocation. $(\mathbf{x^1}; P^1)$ is dominated by $(\mathbf{z^0}; P^0)$ where $\mathbf{z^0} = (2.65, 0.35)$.
- $P^1 \in \mathcal{P}^*$, since $(\mathbf{y^1}; P^1) \in \mathcal{M}^*$ where $\mathbf{y^1} = (1, 2)$. This assertion follows from the fact that $t = 1$ maximizes $3 - t + v(t)$ with value 5. Moreover, $\mathbf{y^1}$ is a P^1-equilibrium allocation.

□□

The scenario of the example is portrayed in the following Fig. 6.1 in terms of utility allocations where U signifies a utility pair (U_1, U_2). The line through $U(\mathbf{x^0})$ and $U(\mathbf{y^0})$ represents the utility pairs at P^0-optimal allocations. The downward sloping part of the curve through $U(\mathbf{y^1})$ and $U(\mathbf{x^1})$ represents the utility pairs at P^1-optimal allocations.

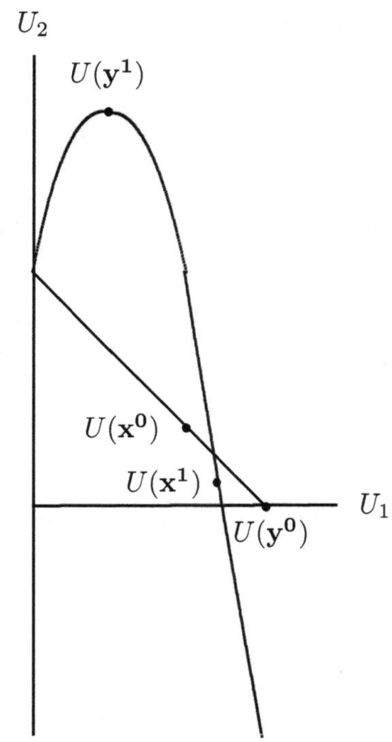

Fig. 6.1 Utility allocations

In the example, there does exist some optimal household structure, P^1 such that $(\mathbf{y^1}; P^1) \in \mathcal{M}^*$ for some P^1-equilibrium allocation, $\mathbf{y^1}$. It remains an open question to what extent this very modest requirement is met beyond the example.

The example demonstrates that the interaction of collective decisions within households and competitive markets can produce inefficient results as soon as the boundaries between microeconomic units are moveable. Namely, suppose that several optimal household structures exist. A competitive equilibrium for a given optimal household structure depends on the collective decisions households undertake and on households' endowments. We have seen that the resulting allocation can be dominated by another optimal household structure together with a feasible reallocation of resources.

The example suggests further that a change in the nature of collective decision making within households may impair household stability. Suppose for example that collective decisions follow cooperative bargaining rules such as the (possibly asymmetric) Nash bargaining solution. Consider now a shift of bargaining power within households. Then a previously optimal allocation of consumers and commodities may be disturbed and one can easily imagine that this brings about a subsequent reshuffling of the sociological structure of the population.

In the example, there are two optimal household structures, P^0 and P^1. In the rare cases with a unique optimal household structure P^*, P^*-equilibrium allocations $\mathbf{x}^*$ yield full Pareto optima $(\mathbf{x}^*; P^*) \in \mathcal{P}^*$:

Proposition 6.2 (Gersbach and Haller (2001), Proposition 5) *Suppose (HSP). Suppose further that $\mathcal{P}^*$ consists of a single element, denoted P^*, and (BE) holds for the economy with household structure P^*. Then every P^*-equilibrium $(p; \mathbf{x})$ satisfies $(\mathbf{x}; P^*) \in \mathcal{M}^*$. In other words, every P^*-equilibrium allocation determines a fully Pareto optimal allocation of commodities and consumers.*

A unique optimal household structure is not required under specific assumptions on preferences:

Proposition 6.3 (Gersbach and Haller (2001), Proposition 4(ii)) *Suppose (PGE), (IPR) and strong monotonicity of consumer preferences. Suppose further that for each consumer i, the utility function $U_i : X_i \to \mathbb{R}$ is concave and in the interior of X_i differentiable. If $P \in \mathcal{P}^*$ and $(p; \mathbf{x})$ is a P^*-equilibrium with $\mathbf{x} \gg 0$, then $(\mathbf{x}; P^*)$ is a fully Pareto optimal allocation.*

The conclusion holds for arbitrary $P \in \mathcal{P}$ if (PGE) is replaced by absence of externalities.

So far, we have dealt with the validity and failure of variations of the first welfare theorem when the household structure is variable. The main message is that even if a social planner can arrange households in an optimal way while the allocation of commodities is left to the market, the resulting overall allocation is not necessarily fully Pareto optimal. The remainder of this chapter is going to be devoted to the second welfare theorem, the implementation of desirable allocations in the sense of Pareto as competitive equilibrium outcomes.

Proposition 6.4 (Second Welfare Theorem) *Suppose $P \in \mathcal{P}$ and that each household $h \in P$ satisfies the redistribution property (RP) and is locally non-satiated. Moreover, for $i \in h \in P$, let the utility function $U_i(\mathbf{x_h}; h)$ be continuous and concave in $\mathbf{x_h} \in \mathcal{X}_h$. If $\mathbf{x}^* \in \mathcal{M}(P)$, $\mathbf{x}^* \gg 0$, and $\omega_h = \sum_{i \in h} x_i^*$ for all $h \in P$, then $\mathbf{x}^*$ is a P-equilibrium allocation.*

The proposition is first stated and proven in Gersbach and Haller (2001, Proposition 6). The proposition asserts that the social planner can decentralize a fully Pareto optimal allocation as a competitive equilibrium among households by rearranging households and by lump-sum redistributions of endowments. The social planner faces, however, the requirement that households conform with his intentions when making their collective decisions. Allocations may be Pareto inefficient when households take another collective decision than envisioned by the social planner, as the next example—also found in Gersbach and Haller (2001)—demonstrates.

Example 6.5 Let $\ell = 1$ and $I = \{1, 2, 3\}$. Consumers $i = 1, 2$ experience no consumption externalities. Consumer 3 is the only person subject to consumption externalities.

Specifically, we assume for $i = 1, 2$ that there exists $u_i : \mathbb{R}_+ \to \mathbb{R}$ with $u_i' > 0$ such that if $i \in h$, $h \in \mathcal{H}$, and $\mathbf{x_h} = (x_j)_{j \in h} \in \mathcal{X}_h$, then

$U_i(\mathbf{x_h}; h) = u_i(x_i)$, in case $h = \{i\}, \{i, 3\}, \{1, 2\}, \{1, 2, 3\}$;
Specifically, we assume for $i = 3$, if $i \in h, h \in \mathcal{H}$, and $\mathbf{x_h} = (x_j)_{j \in h} \in \mathcal{X}_h$:
$U_3(\mathbf{x_h}; h) = x_3$, in case $h = \{3\}$;
$U_3(\mathbf{x_h}; h) = x_3 + v(x_1)$, in case $h = \{1, 3\}$;
$U_3(\mathbf{x_h}; h) = x_3 + v(x_2)$, in case $h = \{2, 3\}$;
$U_3(\mathbf{x_h}; h) = x_3 + v(x_1 + x_2)$, in case $h = I$;
where $v(t) = t - t^2$ for $t \in \mathbb{R}$.

We observe that (BE) is satisfied no matter what the household structure is. We further assume that the social endowment is $\omega_S = 3$ regardless of the household structure.

Suppose that the social planner wants to decentralize the allocation $(\mathbf{x^1}; P^1)$ with $P^1 = \{\{1\}, \{2, 3\}\}$ and $\mathbf{x^1} = (0, 0.5, 2.5)$. Clearly, $(\mathbf{x^1}; P^1) \in \mathcal{M}^*$. Let us pick the desired household structure $P^1 = \{\{1\}, \{2, 3\}\}$ and let lump-sum redistributions of commodities result in zero endowment for household $\{1\}$ and an endowment with three units of the commodity for household $\{2, 3\}$. Suppose now that household $\{2, 3\}$ reaches another efficient collective decision, $\mathbf{z_h} = (\sqrt{3}, 3 - \sqrt{3})$. The resulting P^1-equilibrium allocation $\mathbf{y^1} = (0, \sqrt{3}, 3 - \sqrt{3})$ combined with P^1 lacks full Pareto optimality, since the third consumer suffers a negative consumption externality. $(\mathbf{y^1}; P^1)$ is dominated by $(\mathbf{y^1}; P^2)$ with $P^2 = \{\{1, 3\}, \{2\}\}$, since the third consumer avoids the negative consumption externality when forming a household together with the first consumer. □□

The example demonstrates the limited capability of the social planner to decentralize Pareto efficient allocations. The social planner's possibilities are restricted by the nature of collective decision making of households. The social planner can, however, decentralize any Pareto efficient allocation as a competitive equilibrium among households if he rearranges households and redistributes endowments directly to individuals and if collective decisions within households are individually rational. For instance, individual rationality is satisfied if individuals can use their endowments as status quo points in a cooperative bargaining process.

The limitations of decentralization among households could have important implications for the design of social transfers systems. Relying on efficient choices among households is not sufficient to guarantee a fully Pareto optimal allocation if the recipient of transfers is the household as an entity or one distinguished member of the household. Individualized transfers are necessary to ensure the implementation of socially preferred allocations. However, the policy maker may be frustrated in his efforts to design individualized transfers, if individuals cannot protect their endowments in the collective choice process within households.

The second welfare theorem reminds us of the question of individual decentralization: What would happen if in Proposition 6.4, instead of assigning endowments $\omega_h = \sum_{i \in h} x_i^*$ to all $h \in P$, every person $i \in I$ was given the endowment $\omega_i = x_i^*$ and allowed to shop on her own? We know that as a rule, the outcome of their noncooperative behavior would be inefficient. In exceptional cases, however, individual decentralization yields the targeted Pareto optimum:

Proposition 6.5 *Suppose (PGE) and strong monotonicity of consumer preferences. Suppose further that for each consumer i, the utility function* $U_i : X_i \to \mathbb{R}$ *is concave and in the interior of* X_i *differentiable. If* $P \in \mathcal{P}$ *and* $(p; \mathbf{x})$ *is a P-equilibrium with* $\mathbf{x} \gg 0$*, then* $(p; \mathbf{x})$ *is a competitive equilibrium of the economy where the market participants are individually acting and trading from their consumption bundles* $x_i, i \in I$*.*

The proposition is an immediate consequence of Proposition 2 in Haller (2000).

In general, competitive equilibria cannot be individually decentralized even if they satisfy very stringent normative criteria. To see this, we review and adapt the example of (Gersbach and Haller (2010), p. 721).

Example 6.6 Let $I = \{1, 2\}$, $\ell = 2$, $\omega_1 = \omega_2 = (4, 1)$, $\omega_{\{1,2\}} = \omega_1 + \omega_2 = (8, 2)$. A consumption bundle for individual i is denoted (x_i, y_i) within the example. The specific utility representations are

$$U_i((x_i, y_i); \{i\}) = x_i y_i \text{ for } (x_i, y_i) \in X_i, \text{ i=1,2;}$$

$$U_1((x_1, y_1), (x_2, y_2)); \{1, 2\}) = A x_1 y_1$$
$$\text{for } ((x_1, y_1), (x_2, y_2)) \in X_1 \times X_2;$$

$$U_2((x_1, y_1), (x_2, y_2)); \{1, 2\}) = B x_2 y_2 \cdot (1 + x_1)^{-1}$$
$$\text{for } ((x_1, y_1), (x_2, y_2)) \in X_1 \times X_2,$$

with $A > 3$, $B > 9$.

Now consider the state $(\widehat{p}, \widehat{\mathbf{x}}; \widehat{P})$ given by the price system $\widehat{p} = (\widehat{p}_1, \widehat{p}_2) = (1, 6)$, the commodity allocation $\widehat{\mathbf{x}} = ((\widehat{x}_1, \widehat{y}_1), (\widehat{x}_2, \widehat{y}_2)) = ((2, 1), (6, 1))$, and the household structure $\widehat{P} = \{I\} = \{\{1, 2\}\}$.

Then the following assertions hold:

(i) The state $(\widehat{p}, \widehat{\mathbf{x}}; \widehat{P})$ is a competitive equilibrium at which no group of consumers can benefit from forming a new household.
(ii) The allocation $(\widehat{\mathbf{x}}; \widehat{P})$ is fully Pareto optimal: $(\widehat{\mathbf{x}}; \widehat{P}) \in \mathcal{M}^*$.
(iii) $(\widehat{p}; \widehat{\mathbf{x}})$ is not a competitive equilibrium of the economy where the market participants are individually acting and trading from their consumption bundles $\widehat{x}_i, i \in I$.
(iv) $\widehat{\mathbf{x}}$ cannot be obtained as competitive equilibrium allocation when the market participants are individually acting and trading based on an income sharing rule.

Assertions (i) and (ii) have been shown as CLAIMS 1 and 2 in Gersbach and Haller (2010). Assertions (iii) and (iv) follow from the fact that individual demands are collinear. □□

References

Dufwenberg, M., Heidhues, P., Kirchsteiger, G., Riedel, F., & Sobel, J. (2011). Other-regarding preferences in general equilibrium. *Review of Economic Studies, 78*, 613–639.

Gersbach, H., & Haller, H. (2001). Collective decisions and competitive markets. *Review of Economic Studies, 68*, 347–368.

Gersbach, H., & Haller, H. (2010). Club theory and household formation. *Journal of Mathematical Economics, 46*, 715–724.

Haller, H. (2000). Household decisions and equilibrium efficiency. *International Economic Review, 41*, 835–847.

Mohemkar-Kheirandish, R. (2008). *Intra-household decision making*. Ph.D. thesis, Virginia Polytechnic Institute and State University, Blacksburg, VA.

Chapter 7
General Equilibrium with Endogenous Household Structure

In this chapter, which is primarily based on Gersbach and Haller (2011), we complete the integration of the three previously highlighted allocative mechanisms involving households: (a) Household formation (and dissolution), i.e., individual decisions are made to join or leave households. (b) Household decisions, i.e., collective decisions within households regarding the individual consumption plans of household members. (c) Competitive exchange across households. In the last two chapters, we studied the combination of (b) and (c). Now we add (a), endogenous household formation.

The investigation of endogenous household structures, that is household structures resulting from individual consumer decisions to enter and exit households, presumes variable household structures as already considered in Chap. 6. But now we add specific stability requirements: In equilibrium, consumers no longer wish to change households, to exercise any of the available outside options. We present several equilibrium concepts that differ in the stability requirements for households, in the set of available outside options. We shall focus on two types of outside options available to household members:

(EO) Under the "exit option", a consumer belonging to a multi-member household can decide to leave that household and become a single.

(JO) With the "joining option", an individual can leave its current household—or cease to be single if applicable—and join another already existing household.

Whether an individual is interested in exercising one of its outside options depends on the prevailing household structure and market conditions. It benefits from exercising the exit option if it fares better as an individual consumer at the going prices. The individual has an incentive and opportunity to exercise the joining option if there exist another household and an affordable consumption plan for the augmented household (consisting of the members of the other household plus the joining individual) so that everybody in the augmented household is made better off.

We assume again a finite population $I = \{1, \ldots, N\}$, households $h \in \mathcal{H}$, and household structures $P \in \mathcal{P}$. As before, the commodity space is $\mathbb{R}^\ell$. Consumers have consumption sets $X_i = \mathbb{R}^\ell_+$. Commodity allocations have the form $\mathbf{x} = (x_i)_{i \in I} \in \mathcal{X} = \prod_i X_i$. Overall allocations have the form $(\mathbf{x}; P) \in \mathcal{X} \times \mathcal{P}$. Households have

H. Gersbach and H. Haller, *Groups and Markets*,
DOI 10.1007/978-3-319-60516-6_7

endowments ω_h. If household structure P prevails, then the social endowment is $\omega_P = \sum_{h \in P} \omega_h$ and feasibility of an allocation $(\mathbf{x}; P)$ means $\sum_i x_i = \omega_P$. For the remainder of this chapter, we assume IPR so that $\omega_h = \sum_{i \in h} \omega_{\{i\}}$ for all $h \in \mathcal{H}$ and the social endowment is thus independent of P and denoted ω_S.

We maintain the **Assumption of Household-Specific Preferences (HSP)** throughout this chapter, with individual utility representations $U_i : \mathcal{A}_i \to \mathbb{R}$ where $\mathcal{X}^* = \bigcup_{h \in \mathcal{H}} \mathcal{X}_h$ and $\mathcal{A}_i = \{(\mathbf{x_h}; h) \in \mathcal{X}^* \times \mathcal{H} : h \in \mathcal{H}_i, \mathbf{x_h} \in \mathcal{X}_h\}$ for $i \in I$. Additional notation like $P(i), \mathcal{H}_i, \mathcal{X}(P), \mathcal{M}(P), \mathcal{M}^*$ is defined as before.

Further define a **state** of the economy as a triple $(p, \mathbf{x}; P)$ such that $p \in \mathbb{R}^\ell$ is a price system and $(\mathbf{x}; P) \in \mathcal{X} \times P$ is an allocation, i.e., $\mathbf{x} = (x_i)_{i \in I}$ is an allocation of commodities and P is an allocation of consumers (a household structure, a partition of the population into households). We say that in state $(p, \mathbf{x}; P)$,

(a) consumer i can benefit from exit, if $P(i) \neq \{i\}$ and there exists $y_i \in B_{\{i\}}(p)$ such that $U_i(y_i; \{i\}) > U_i(\mathbf{x_{P(i)}}; P(i))$;
(b) consumer i can benefit from joining another household g, if $g \in P$, $g \neq P(i)$ and there exists $\mathbf{y_{g \cup \{i\}}} \in B_{g \cup \{i\}}(p)$ such that $U_j(\mathbf{y_{g \cup \{i\}}}; \mathbf{g} \cup \{\mathbf{i}\}) > U_j(\mathbf{x_{P(j)}}; P(j))$ for all $j \in \mathbf{g} \cup \{\mathbf{i}\}$.
(c) a group of consumers h can benefit from forming a new household, if $h \notin P$ and there exists $\mathbf{y_h} \in B_h(p)$ such that $U_j(\mathbf{y_h}; h) > U_j(\mathbf{x_{P(j)}}; P(j))$ for all $j \in h$.

Equilibrium Concepts. We amend the definition of a competitive equilibrium among households in two ways, based on the availability of outside options. We first introduce the concept of a **competitive equilibrium with free exit (CEFE)** which is a competitive equilibrium where no individual can benefit from exercising the exit option. Second, we introduce the concept of a **competitive equilibrium with free household formation (CEFH)** which is a competitive equilibrium where no individual can gain from exercising the exit option or joining another household. While the focus of our analysis lies on the two equilibrium concepts CEFE and CEFH, we shall also consider competitive equilibria where no group of consumers can benefit from forming a new household. That individual consumers or groups of consumers should not benefit from leaving, switching groups or forming new groups are stability requirements known from the literatures on matching and clubs. We are going to investigate how more stringent equilibrium conditions in the form of more demanding stability requirements affect equilibrium existence and equilibrium welfare.

Informally, we define an equilibrium of commodities and consumers as a price system together with a household structure and a feasible resource allocation such that:

- a household chooses an efficient consumption schedule for its members, subject to the household budget constraint;
- markets clear;

- no individual has an incentive to leave a household and to participate as an individual in the market at the going prices. That is, no individual would benefit from exercising the exit option. What it means to benefit from exit is formalized under (a) above.

These three conditions define a **competitive equilibrium with free exit**. We shall consider a fourth condition, that

- no individual can leave a household and get accepted by another household by proposing a feasible allocation for the enlarged household which makes everybody in this newly formed household better off at the going prices. That is, no individual would benefit from exercising the joining option. What it means to benefit from joining another household is formalized under (b) above.

The combination of all conditions defines a **competitive equilibrium with free household formation**. Formally, we define the equilibrium concepts as follows:

Definition A state $(p, \mathbf{x}; P)$ is a **competitive equilibrium (among households)** if it satisfies the following conditions:

1. $\mathbf{x_h} \in EB_h(p)$ for all $h \in P$.
2. $\sum_i x_i = \omega_S$.

A state $(p, \mathbf{x}; P)$ is a **competitive equilibrium with free exit (CEFE)** if it satisfies conditions 1, 2 and

3. No consumer can benefit from exit.

A state $(p, \mathbf{x}; P)$ is a **competitive equilibrium with free household formation (CEFH)** if it satisfies conditions 1, 2, 3, and

4. No consumer can benefit from joining another household.

Discussion. Condition 1 reflects collective rationality in contrast to the traditional "unitary" model where households are treated like single consumers. Efficient choice by the household refers to the individual consumption and welfare of its members, not merely to the aggregate consumption bundle of the household. Condition 2 requires market clearing. Conditions 1 and 2 alone also define a **competitive equilibrium** $(p, \mathbf{x})$, given household structure P, previously called a P-equilibrium. In fact, the state $(p, \mathbf{x}; P)$ is a competitive equilibrium (among households) if and only if $(p; \mathbf{x})$ is a P-equilibrium.

Condition 3 constitutes an individual rationality or voluntary participation (membership) constraint. Conditions 1 to 3 together define a **competitive equilibrium with free exit (CEFE)**.

Conditions 1 to 4 together define a **competitive equilibrium with free household formation (CEFH)**. Condition 4 stipulates that no individual can leave a household and can propose a feasible consumption allocation to the members of a new household, created by the individual and another already existing household, which makes everybody in the new household better off at the going prices. A stronger condition rules out that all members of the new household are made weakly better off (and the moving person strictly better off) at the prevailing prices. That is, it rules out

(b′) existence of $i \in I, g \in P, g \neq P(i)$, and $\mathbf{y}_{\mathbf{g}\cup\{\mathbf{i}\}} \in B_{g\cup\{i\}}(p)$ such that
$U_j(\mathbf{y}_{\mathbf{g}\cup\{\mathbf{i}\}}; g \cup \{i\}) \geq U_j(\mathbf{x_g}; g)$ for all $j \in g$;
$U_i(\mathbf{y}_{\mathbf{g}\cup\{\mathbf{i}\}}; g \cup \{i\}) > U_i(\mathbf{x_h}; h)$.

All our results continue to hold with the stronger requirement. Here we adopt the weaker condition 4 because it facilitates comparisons with the literature and some of our own work.

One could imagine even stronger equilibrium conditions in the tradition of the matching literature (see Roth and Sotomayor (1990) for surveys) requiring, for instance, that two persons cannot benefit from breaking away from two different matches and forming a new match. The condition that no group of consumers can benefit from forming a new household encompasses the latter requirement as well as conditions 3 and 4. On the contrary, it has been argued in other contexts that the divorce threat and thus the exit option alone describes the behavior of individuals in multi-person households; see Bourguignon and Chiappori (1994) for a summary of this debate. The combination of our conditions 3 and 4 lies between these two perspectives of how individuals decide whether to leave a household. It proves sufficient to put the existence of equilibria into question; see Examples 7.1 and 9.1 for non-existence of equilibria with free household formation. It is just restrictive enough to make the normative issue how more outside options affect welfare an interesting one; see Sect. 7.3. Thus the focus of our analysis lies on CEFE and CEFH—though we shall invoke at times the more demanding concept of a competitive equilibrium at which no group of consumers benefits from forming a new household. Note that CEFH may be quite appropriate in the context of firms. For instance, when firms are governed by partnership agreements, partners may contemplate switching firms. It ultimately depends on social and legal norms which outside options are available to or contemplated by individuals. A change of norms can alter the set of available outside options. In turn, the stability of particular households and the welfare of certain individuals can be affected.

Equilibrium outcomes are not affected by the qualifications CEFE and CEFH in the absence of externalities. The following neutrality theorem has been demonstrated in Gersbach and Haller (2003) and also reported in Gersbach and Haller (2011).

Proposition 7.1 (Neutrality Theorem)
Suppose absence of externalities and continuity and local non-satiation of consumer preferences. Consider $(p; \mathbf{x}) \in \mathbb{R}^{\ell} \times \mathcal{X}$ and any household structure P. Then the following three assertions are equivalent:

(i) $(p, \mathbf{x}; P)$ is a competitive equilibrium with free household formation.

(ii) $(p, \mathbf{x};\ P)$ *is a competitive equilibrium with free exit.*
(iii) $(p, \mathbf{x})$ *is a traditional competitive equilibrium where each agent acts and trades individually.*

Proposition 7.1 conforms with intuition. It asserts that in the absence of any externalities, free exit implies that a consumer can fare no better and no worse as a member of a multi-member household than as an individual market participant. If, in spite of free exit, some individuals enjoy higher utility levels as household members than they would obtain individually, some sort of externality has to be present.

7.1 Existence of Equilibria with the Exit Option

In this section we establish the existence of competitive equilibria with free exit (CEFE).

Trivial Equilibria. We first observe that for the household structure where all households are singletons, denoted by $P^0 = \{\{1\}, \ldots, \{n\}\}$, there frequently exists a trivial competitive equilibrium with free exit of the form $(p, \mathbf{x};\ P^0)$. This is an immediate corollary of the proof of Proposition 7.2 below, which is given in the Appendix of Gersbach and Haller (2011). In that proof, we obtain existence of a price system p and an allocation $\mathbf{x}$ so that conditions 1 and 2 for a competitive equilibrium with free exit are satisfied. We need not check condition 3, since all individuals are already in one-person households which renders the exit option obsolete.

We have just established existence of trivial competitive equilibria with exit where everybody is single and not exposed to externalities. We also know that under the provisions of the neutrality theorem, any household structure qualifies as equilibrium household structure, provided there is an equilibrium at all. Otherwise, for a multi-member household to exist in equilibrium, there ought to be some incentive for multi-member household formation, some advantage (from living in a larger household) that prevents its members from leaving.

Large Group Advantage and Non-trivial Equilibria: Motivation. A priori, a large group or, to be precise, a non-single household h offers an advantage to its members if at any given price system, the group can afford consumption plans for its members that make each member better off than the member's optimal choice as a single consumer—which is captured by inequalities of the form (7.1) below.

If preferences are assumed convex and continuous in household consumption, then under certain additional assumptions, Debreu's (1952) social equilibrium approach to the equilibrium existence problem proves most suitable. One crucial step in the social equilibrium approach is the restriction to truncated budget sets. This technicality makes the formal definition of the Large Group Advantage (LGA) condition below more complicated and elaborate, since one has to make sure that the inequalities (7.1) can be met if household h is restricted to a truncated budget set. Formally, this requirement is captured by the following conditions 1 and 2. To this end, we restrict prices to the price simplex

$$\Delta = \left\{ p \in \mathbb{R}_+^\ell : \sum_{k=1}^{\ell} p^k = 1 \right\}.$$

We denote the relative interior of Δ by $\mathring{\Delta}$. Further let us choose $k > 0$ so that the social endowment ω_S belongs to the cube $Q = [0, k]^\ell$. Set $K = [0, 2k]^\ell$.

(LGA) Large Group Advantage: *We say that a multi-member household h has large group advantage, if:*

1. *Every member $i \in h$ has a utility maximizer $x_i^0(p)$ and achieves indirect utility $v_i(p) = U_i(x_i^0(p); \{i\})$ when trading individually from the endowment $\omega_{\{i\}}$ at prices $p \in \mathring{\Delta}$.*
2. *For every price system $p \in \Delta$, there exists a set $X_h(p) \subseteq B_h(p) \cap K^h$ with the following properties:*

2.a. *$X_h(p)$ is non-empty, compact and convex.*
2.b. *$X_h(p)$ depends continuously on p.*
2.c. *There exist thresholds $\delta_i(p) \geq 0$ for $p \in \mathring{\Delta}$ and $i \in h$ such that for $p \in \mathring{\Delta}$ and $\mathbf{x_h} \in B_h(p) \cap K^h$: $\mathbf{x_h} \in X_h(p)$ iff*

$$U_i(\mathbf{x_h}; h) - v_i(p) \geq \delta_i(p) \tag{7.1}$$

holds for all $i \in h$.

Remark 7.1 To illustrate that the key condition (2) of LGA is non-vacuous, let us present two alternative assumptions on a multi-member household h that will yield condition (2) when supplemented with suitable further assumptions: (i) Sufficiently bounded individual demands so that $(x_i^0(p))_{i \in h} \in B_h(p) \cap K^h$. An example is given by the utility representation $U_i(x_i; \{i\}) = \min\{x_i^k | k = 1, \ldots, \ell\}$. (ii) Group preferences which strictly dominate consumption preferences, e.g. $U_i(\cdot\,; h) \geq 0$ and $U_i(\cdot\,; \{i\}) < 0$ for $i \in h$. In this case, (7.1) becomes trivial. An example with $U_i(\cdot\,; \{i\}) < 0$ is given by $U_i(x_i; \{i\}) = -\sum_k \exp(-x_i^k)$. The somewhat extreme cases (i) and (ii) have the virtue of being simple and transparent.

Remark 7.2 The question remains whether one could not simply set $\delta_i \equiv 0$ for all i and make standard assumptions on preferences that imply LGA. Indeed, if for each $i \in h$, the functions $U_i(\cdot\,; \{i\})$ and $U_i(\cdot\,; h)$ are continuous and concave, then:

- $v_i(\cdot)$ is well defined and continuous;
- and for $p \in \mathring{\Delta}$,

$$X_h(p) \equiv \{\mathbf{x_h} \in B_h(p) \cap K^h | \, U_i(\mathbf{x_h}; h) \geq v_i(p) \; \forall i \in h\}$$

is a compact and convex set that depends continuously on p. However, LGA stipulates in addition: (a) $X_h(p) \neq \emptyset$ and (b) existence of a continuous extension of the correspondence $X_h(\cdot)$ from $\mathring{\Delta}$ to Δ with the desired properties. With $\delta_i \equiv 0$ for all i, (a) has the best chance to be met whereas the further requirement (b) may not be satisfied, but might hold for some $\delta_i(p) > 0$. Hence imposing (2) with arbitrary $\delta_i(p) \geq 0$ is more general than setting $\delta_i(p) \equiv 0$.

We obtain:

Proposition 7.2 (Non-Trivial Equilibria; Proposition 2 of Gersbach and Haller (2011))
Suppose:

(i) $\omega_h \gg 0$ *for all* $h \in \mathcal{H}$.
(ii) $U_i(\mathbf{x_h}; h)$ *is continuous and concave for all* $i \in h, h \in \mathcal{H}$.
(iii) $U_i(x_i; \{i\})$ *is strictly monotone for all* $i \in I$.
(iv) There exist a household $h \in \mathcal{H}$ *with* $1 < |h| < n$, *which has large group advantage (LGA), and a member* $j \in h$ *whose preferences are strictly monotonic in own consumption and who is not imposing any negative consumption externalities on other household members.*

Then there exists a competitive equilibrium with free exit of the form $(p, \mathbf{x}; P)$ *with* $P \neq P^0$. *More specifically,* $h \in P$ *for some* h *satisfying (iv).*

The informal idea of the proposition is that as soon as two or more agents can gain from living together in a household as stipulated by LGA, there exist non-trivial equilibria with free exit where this particular multi-member household is formed.

Remark 7.3 There are two *immediate extensions*: First, it is straightforward to extend Proposition 7.2 to equilibria with free exit, at least one single-person household and several multi-member households if LGA holds for each of these multi-member households. Second, in the proof, demand for the multi-member household at hand is derived from the maximization of an equally weighted sum of household members' utilities. Closer inspection of the proof indicates that it also works with any unequal positive weights, i.e., with household h maximizing $W_h = \sum_{i \in h} \lambda_i U_i(\mathbf{x_h}; h)$ and all $\lambda_i > 0$. In fact, any continuous and concave Paretian social welfare function for the household will do, for instance $W_h(\mathbf{x_h}) = -\sum_{i \in h} \exp[-U_i(\mathbf{x_h}; h)]$.

Remark 7.4 With different assumptions, an approach relying on Lemma 4.1, the "excess demand lemma", proves successful. Specifically, one makes certain standard assumptions (including strict concavity of the functions $U_i(\cdot\,; \{i\})$) in combination with particular positive externalities of the separable type within household h. Two special cases of the latter are positive pure group externalities [(PGE) restricted to household h] on the one hand and positive separable pure consumption externalities [(SEP) of Haller (2000) restricted to household h] on the other hand. The proof is similar to that of Proposition 3 in Gersbach and Haller (1999); see also Proposition 4.5. Without separability and purely positive externalities, the social equilibrium approach of Debreu taken above is more promising.

7.2 Existence of Equilibria with the Exit and the Joining Option

In this section, we take up the challenging question whether and under which circumstances competitive equilibria with free household formation (CEFH) exist. We start with an existence result, followed by a non-existence result and further counter-examples.

Proposition 7.3 (Gersbach and Haller (2011), Proposition 3) *Suppose pure group externalities, that is*
$U_i(\mathbf{x_h}; h) = U_i^c(x_i) + U_i^g(h)$ *for* $\mathbf{x_h} \in \mathcal{X}_h, h \in \mathcal{H}_i, i \in I$. *If*

(i) $\omega_S \gg 0$, *each of the functions* $U_i^c, i \in I$, *is continuous, strictly increasing and strictly quasi-concave, and*
(ii) there is a unique optimal household structure P based solely on group preferences represented by $U_i^g, i \in I$,

then a fully Pareto-optimal CEFH with equilibrium household structure P exists.

It turns out that condition (ii) cannot be dispensed with:

Proposition 7.4 (Gersbach and Haller (2011), Proposition 4) *There are economies where pure group externalities of the form* $U_i(\mathbf{x_h}; h) = U_i^c(x_i) + U_i^g(h)$ *for* $\mathbf{x_h} \in \mathcal{X}_h, h \in \mathcal{H}_i, i \in I$ *prevail; condition (i) of Proposition 7.3 is satisfied; a CEFE exists; and a CEFH does not exist.*

Proposition 7.4 suggests that CEFH—and a fortiori competitive equilibria at which no group benefits from forming a new household—do not exist in general. Next we perform a more detailed examination of instances where competitive equilibria with free household formation need not exist, where all conceivable household structures are destabilized by outside options of the second type (JO).

Example 7.1 Let $I = \{1, 2, 3, 4\}$ and $\ell = 1$. For a household h, the endowment is $w_h = |h|$. Preferences are represented by utility functions $U_i, i \in I$, and given as follows:

$$U_i(\mathbf{x_h}; h) = u(x_i) \quad \text{if } h = \{i\} \tag{7.2}$$
$$U_i(\mathbf{x_h}; h) = u(x_i) + k \quad \text{if } |h| = 2 \tag{7.3}$$
$$U_i(\mathbf{x_h}; h) = u(x_i) + k \quad \text{if } |h| = 3,\ i = 1, 2 \tag{7.4}$$
$$U_i(\mathbf{x_h}; h) = u(x_i) + k + \varepsilon \quad \text{if } |h| = 3,\ i = 3, 4 \tag{7.5}$$
$$U_i(\mathbf{x_h}; h) = u(x_i) - k \quad \text{if } |h| = 4 \tag{7.6}$$

The group externalities satisfy $k > 0$ and $k \geq \varepsilon \geq 0$. The function u is continuous and strictly increasing. It satisfies $u(1) \geq u(0) + k$. Since there is only one good, we can set $p = 1$.

We first consider the case $\varepsilon = 0$. Then, there exists a CEFH, namely

$$E^1 = \big(p_;(1,1,1,1)_;\{\{1,2\},\{3,4\}\}\big)$$

with utility allocation $(u(1)+k, u(1)+k, u(1)+k, u(1)+k)$. Since the population is homogeneous, there exist two other equilibria with the same utility allocation and household structures $\{\{1,3\},\{2,4\}\}$ and $\{\{1,4\},\{2,3\}\}$, respectively. No other equilibria with free household formation exist. For instance, the household structure $\{\{1,2,3\},4\}$ cannot be part of an equilibrium, since at least one individual in the household $\{1,2,3\}$ can propose to agent $i = 4$ to form a two-person household which makes both individuals better off. Specifically, the individual leaving $\{1,2,3\}$ can offer $i = 4$ a consumption level $u^{-1}(u(1) - k + \delta)$ for some small $\delta, k > \delta > 0$. Agent 4's utility will be $u(1) + \delta$ and therefore larger than in the candidate equilibrium. The deviating agent obtains a utility

$$u(2 - u^{-1}(u(1) - k + \delta)) + k$$

which exceeds the utility of at least one member in the household $\{1,2,3\}$ since $\delta < k$.

Next let us consider the case $\varepsilon > 0$ where ε is sufficiently small. We claim that no CEFH exists. Consider first the candidate equilibrium E^1. Individual 2 could join $\{3,4\}$ by proposing the household allocation:

$$x_g = (x_2, x_3, x_4) = \big(3 - 2u^{-1}(u(1) - \varepsilon), u^{-1}(u(1) - \varepsilon), u^{-1}(u(1) - \varepsilon)\big) \quad (7.7)$$

which yields the utility allocation

$$\big(u(3 - 2u^{-1}(u(1) - \varepsilon)) + k, u(1) + k, u(1) + k\big) \quad (7.8)$$

and makes agent 2 better off while the utility of individuals 3 and 4 remains constant. Hence, E^1 cannot be a CEFH. A similar argument applies mutatis mutandis for any other household structure with two-person households. Furthermore, by essentially the same argument as before, no CEFH can exist with a three-person or four-person household. Finally, if everybody were alone, two persons could form a household and both be better off. Therefore, no CEFH exists. □□

The interesting feature of the example is that a small change of the externalities destroys the existence of a competitive equilibrium with free household formation. It is obvious that the existence problem in the example can be overcome by taking a specific number of replica of the original economy. In the example three replica would allow all individuals preferring a three-person household over a two-person household to be member of a three-person household while other individuals could live in two-person households.

However, enlarging the economy through replication cannot restore existence in other examples. The most intriguing of those examples is Example 3 of Gersbach and Haller (2011), reproduced as Example 9.1. It is of the matching type. That is, preferences are such that the only single households and heterosexual couples can emerge in equilibrium. Yet because of particular externalities and active commodity markets, the classical existence of stable matchings no longer holds. There are two commodities in the example, one female called Anita, and two ex ante identical males, say Peter and Paul. It is always Anita who wants to change households. Suppose Anita and Peter form a household while Paul is single. With this household structure in place, markets are cleared at a price system p^*. Then, however, Anita finds it optimal to split at the going market prices in order to reduce negative consumption externalities. But at the market clearing price system p^{**} associated with the new household structure, Anita (and Peter or Paul) find it optimal to form a two-person household in order to benefit from group externalities, because they can buy more of the goods which generate less consumption externalities. Actually, in the best outcome for Anita, the male partner is indifferent between remaining single and being matched. But now the market clearing price system becomes p^* again and, consequently, it is better for Anita to go single again. The marital status of the woman in the example affects her market opportunities and vice versa. It is impossible to obtain a stable matching and market clearing simultaneously. For certain parameter configuations, this impossibility persists under replication. The basic pattern is depicted in Fig. 7.1.

The example illustrates why active trade across households poses a serious challenge for the traditional matching literature. Given a price system p, one can assign to every single household or couple h a set of affordable utility allocations denoted $V_h(p)$. Given the family of sets $V_h(p)$, one can possibly impose suitable restrictions and obtain existence of a "stable matching" P. Let $\mathcal{S}(p) \subseteq \mathcal{P}$ denote the set of all those stable matchings. Conversely, given a household structure P, one may find a market clearing price system under certain conditions. Let $\mathfrak{P}(P)$ denote the set of all market clearing price systems given P. The problem is to find a fixed point of the correspondence $p \mapsto \mathfrak{P}(\mathcal{S}(p))$. The example demonstrates that such a fixed point need not exist: For any stable matching P given a price system p, the price system p need not be market clearing given the matching (household structure) P.

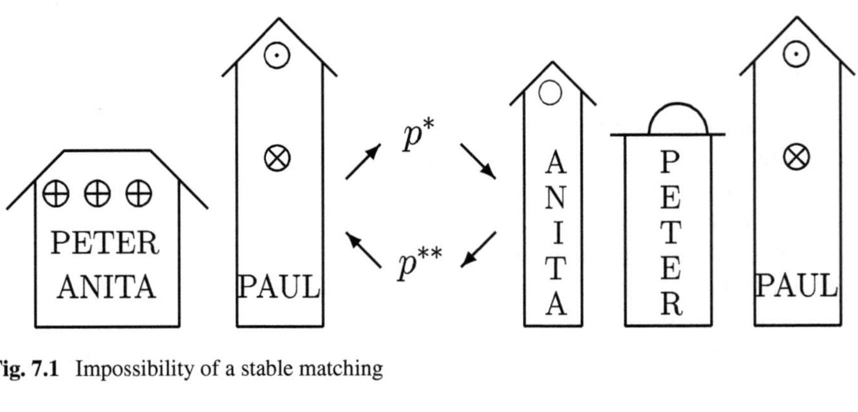

Fig. 7.1 Impossibility of a stable matching

7.3 Equilibrium Welfare

We have already defined a **fully or strongly Pareto-optimal** allocation as a feasible allocation $(\mathbf{x}; P)$ such that there is no feasible allocation $(\mathbf{y}; Q)$ satisfying $(U_i(y_{Q(i)}; Q(i)))_{i\in I} > (U_i(x_{P(i)}; P(i)))_{i\in I}$. Three more definitions prove useful to state welfare properties of competitive equilibria among households. A feasible allocation $(\mathbf{x}; P)$ is **weakly Pareto-optimal**, if there is no feasible allocation $(\mathbf{y}; Q)$ such that $U_i(y_{Q(i)}; Q(i)) > U_i(x_{P(i)}; P(i))$ for all $i \in I$. A feasible allocation $(\mathbf{x}; P)$ is a **weak core allocation**, if there do not exist a non-empty subset J of I, a partition Q of J into households and consumption bundles $y_j \in X_j$ for $j \in J$ such that $\sum_{j\in J} y_j = \sum_{h\in Q} \omega_h$ and $U_j(y_{Q(j)}; Q(j)) > U_j(x_{P(j)}; P(j))$ for all $j \in J$. $(\mathbf{x}; P)$ is a **strong core allocation**, if there do not exist $\emptyset \neq J \subseteq I$, a partition Q of J and consumption bundles $y_j \in X_j$ for $j \in J$ such that $\sum_{j\in J} y_j = \sum_{h\in Q} \omega_h$, and $U_j(y_{Q(j)}; Q(j)) \geq U_j(x_{P(j)}; P(j))$ for all $j \in J$, and $U_j(y_{Q(j)}; Q(j)) > U_j(x_{P(j)}; P(j))$ for some $j \in J$.

Strong equilibrium conditions yield strong welfare properties, namely weak core inclusion:

Proposition 7.5 (Gersbach and Haller (2005), Proposition 4)
Let $(p, \mathbf{x}; P)$ be a competitive equilibrium at which no group benefits from forming a new household. Then $(\mathbf{x}; P)$ belongs to the weak core and, consequently, is weakly Pareto-optimal.

If one assumes in addition the budget exhaustion property and the redistribution property of Gersbach and Haller (2001), then weak core inclusion can be replaced by strong core inclusion.

There are two reasons to explore the welfare properties of the less demanding equilibrium concepts CEFE and CEHF as well. First, we can show existence of CEFH only under the assumption of pure group externalities. In particular, we obtain an example (described in the last section) of non-existence in a two-sided matching model with two goods and consumption externalities. The more demanding requirement that no group benefits from forming a new household renders existence even less likely. Second, we can compare the welfare properties of CEFH and of CEFE which are not CEFH. As a rule, CEFH constitutes a more stringent equilibrium concept than CEFE. One might conjecture that more stringent equilibrium conditions make the surviving equilibria "stronger" or "better", having passed more tests than the eliminated ones. It turns out that this conclusion is not true if "better" means "Pareto-superior".

Optimality of CEFE. It is obvious that competitive equilibrium allocations with free exit need not be fully Pareto-optimal. Suppose, e.g., that there are large gains from forming a two-person household because two individuals, say consumers 1 and 2, have positive pure group externalities. No further externalities are present in the economy. Moreover, suppose that both consumers have the same endowments and the same consumption preferences. A competitive equilibrium with free exit can have every person live in a single-person household. This equilibrium is, however, not Pareto-optimal. Consumers 1 and 2 could form a two-person household with a household excess demand function equal to the sum of individual excess demand functions. Hence, equilibrium prices and consumption allocation would remain as if all persons lived in single-person households. Hence, consumers 1 and 2 would be better off while all other individuals receive the same utility. The example suggests that the lack of appropriate outside options causes inefficiency of CEFE. It also demonstrates that a pair of CEFE may be Pareto-rankable. In the next paragraph we discuss CEFH.

Welfare Implications of Adding the Joining Option. Proposition 7.1 identifies conditions under which adding outside options proves irrelevant if there are no externalities. Now we are going to examine the consequences of providing more outside options to household members in the presence of externalities. Clearly the additional requirement, that no consumer can benefit from joining another household, can eliminate some of the competitive equilibria with free exit. But which ones? The good ones, the bad ones, all or none? We shall demonstrate by means of examples that each of the four conceivable alternatives is possible, indeed.

We have already seen that under the hypothesis of Proposition 7.1, none of the equilibria is eliminated. Example 7.1 constitutes a case where all equilibria are eliminated. In particular, Example 7.1 exhibits existence of CEFE and non-existence of CEFH; see Proposition 4 of Gersbach and Haller (2011) and its proof. In the following, we will demonstrate the other two possibilities. Let us first examine an example that exhibits a pair of weakly Pareto-rankable competitive equilibria with free exit where the inferior one is also a competitive equilibrium with free household formation whereas the superior one is not. Subsequently, we modify the example so that the superior competitive equilibrium with free exit turns out to be a competitive equilibrium with free household formation while the inferior equilibrium is eliminated by the additional requirement.

In both examples, the prospect of a tiny surplus share induces a currently single person to form a two-person household, leaving most of the surplus to the new partner. With a population of three people, this leads to the break-up of any existing two-person household and formation of a new one if the opportunity arises, that is if the joining option becomes available. In contrast, the members of a three-person household have no other household to join in a population of three people; thus the three-person household remains unaffected by the introduction of the joining option. In the examples one can suitably alter the household structure associated with a Pareto-superior competitive equilibrium with free exit by varying (primarily) the per capita surplus in three-person households.

Example 7.2 Let $I = \{1, 2, 3\}$ and $\ell = 1$. For a household h, let the endowment be $\omega_h = |h|$. Let preferences have utility representations of the form

$$U_i(\mathbf{x_h}; h) = a(|h|) \cdot x_i$$

for consumer i in household h where $a(1) = 2, a(2) = 8, a(3) = 5$. Since there is only one good and preferences are strictly monotone, we can set $p = 1$. First consider the competitive equilibrium with free exit $E^1 = (p; (1, 1, 1); \{\{1\}, \{2, 3\}\})$ with utility allocation $(2, 8, 8)$. Next consider the CEFE $E^2 = (p; (0.4, 1.3, 1.3); \{I\})$ with utility allocation $(2, 6.5, 6.5)$. Then E^1 weakly Pareto-dominates E^2. The inferior equilibrium is also a CEFH, since there is no other household to join. However, the superior equilibrium is not a CEFH. Namely individual 2 can propose to consumer 1 to form household $\{1, 2\}$ with consumption $y_1 = 1/2,\ y_2 = 3/2$ which makes both better off. □□

Example 7.3 Let again $I = \{1, 2, 3\}$ and $\ell = 1$. Modify the previous example by setting $a(1) = 1, a(2) = 8, a(3) = 6$. Take E^1 as before, now with utility allocation $(1, 8, 8)$. Set $E^2 = (p; (1/5, 7/5, 7/5); \{I\})$ which is an efficient CEFH, with utility allocation $(1.2, 8.4, 8.4)$. Here E^1 is a CEFE, is strictly dominated by E^2 and is not a CEFH. □□

7.4 Outlook

We have developed a general equilibrium model where households operate in a competitive market environment, can have several members and make efficient collective consumption decisions. Our approach differs from partial equilibrium analysis which offers many valuable insights into household-related issues, but misses the feedback between household formation and household decisions on the one hand and competitive market conditions, in particular price formation on the other hand. We trust that applications of our models can provide a complimentary general equilibrium perspective on these issues. Specific applications are presented in Chaps. 12, 14 and 15.

References

Bourguignon, F., & Chiappori, P.-A. (1994). The collective approach to household behavior. In R. Blundell, I. Preston, & I. Walker (Eds.), *The measurement of household welfare* (pp. 70–85). Cambridge, UK: Cambridge University Press.

Debreu, G. (1952). A social equilibrium existence theorem. *Proceedings of the National Academy of Sciences USA*, *38*, 886–893.

Gersbach, H., & Haller, H. (1999). Allocation among multi-member households: Issues, cores and equilibria. In A. Alkan, C. D. Aliprantis, & N. C. Yannelis (Eds.), *Current trends in economics: Theory and applications*. Heidelberg: Springer-Verlag, Berlin.

Gersbach, H., & Haller, H. (2001). Collective decisions and competitive markets. *Review of Economic Studies*, *68*, 347–368.

Gersbach, H., & Haller, H. (2003). Competitive markets, collective decisions and group formation. CESifo Working Paper 953.

Gersbach, H., & Haller, H. (2005). When inefficiency begets efficiency. *Economic Theory*, *25*, 105–121.

Gersbach, H., & Haller, H. (2011). Competitive markets, collective decisions and group formation. *Journal of Economic Theory*, *146*, 275–299.
Haller, H. (2000). Household decisions and equilibrium efficiency. *International Economic Review*, *41*, 835–847.
Roth, A. E., & Sotomayor, M. A. O. (1990). *Two-sided matching: A study in game-theoretic modeling and analysis*. Cambridge, UK: Cambridge University Press.

Chapter 8
Cores

This chapter primarily builds upon the treatment of cores in Gersbach and Haller (1999). Like Pareto optimal allocations, core allocations can be defined without recourse to market prices. Which allocations qualify as core allocations depends on which coalitions can form and what these coalitions can achieve on their own. For the purpose of core theory, we interpret the set $\mathcal{H} = \{h \subseteq I | h \neq \emptyset\}$ in a dual way: as the set of all potential households and as the set of potential coalitions. Furthermore, let $\mathcal{F}$ denote the set of feasible allocations of the economy. Following Haller (1990), we consider an improvement relation $\mathbb{I}$ from $\mathcal{H}$ to $\mathcal{F}$, that is, $\mathbb{I} \subseteq \mathcal{H} \times \mathcal{F}$. If $(S, F) \in \mathbb{I}$, we say that *coalition S can improve upon allocation F*. (The earlier expression *coalition S can block F* has fallen in disgrace and been replaced by *coalition S can improve upon F*). In a traditional pure exchange economy this may mean, for instance, that the members of S can trade their resources among themselves in such a way that each member ends up with a consumption bundle he prefers to the one F assigns to him. The improvement relation $\mathbb{I}$ may impose restrictions on coalitions. For example, $(S, F) \in \mathbb{I}$ may require that S is contained in a connected component of a given social network in the population I. In an economy with production and private property, an improvement may necessitate a change of production plans and a coalition would have to have control of the firms whose production plans it wants to alter. An improvement relation $\mathbb{I}$ defines a core $C_{\mathbb{I}} = \{F \in \mathcal{F} :$ There is no $S \in \mathcal{H}$ such that $(S, F) \in \mathbb{I}.\}$. In other words, $C_{\mathbb{I}}$ consists of the feasible allocations no coalition can improve upon.

In the traditional model of pure exchange, a coalition is treated as a sub-economy. We adhere to this tradition, while allowing for multi-member households. Then feasible allocations for a coalition or sub-economy are defined in analogy to feasible coalitions for the entire economy. Thus within a coalition or sub-economy, like in the full economy, households form and the available resources are allocated. This means that a coalition has to come up with a household structure of its own, a non-traditional, but quite natural requirement. It is a mild consistency condition, presuming that coalitions are subject to similar constraints as society at large. Consequently,

H. Gersbach and H. Haller, *Groups and Markets*,
DOI 10.1007/978-3-319-60516-6_8

if the household structure is variable, then a coalition should also be flexible in its choice of household structure. If on the other hand, the population has to live with a fixed household structure, then every coalition ought to conform with this household structure as well. The second requirement that a coalition has to rely on its own resources (whose quantity may depend on its choice of household structure) and to allocate those among its members is standard. Depending on whether the household structure is fixed or not, we distinguish between two core concepts, a constrained or household core and the full core.

Let us first formalize the feasibility conditions for a coalition S. Let $\mathcal{Q}_S$ denote the collection of partitions of S into non-empty subsets. An element $Q \in \mathcal{Q}_S$ represents a **household structure for coalition** S. The resources available to coalition S with household structure $Q \in \mathcal{Q}_S$ equal $\omega^*_Q = \sum_{h \in Q} \omega_h$. S is only a sub-economy in its own right if it allocates consumers and commodities under **autarky**. Accordingly, we define a feasible allocation for S as a pair $(\mathbf{y}_S; Q) \in \mathcal{X}_S \times \mathcal{Q}_S$ such that $\overline{\mathbf{y}}_S = \omega^*_Q$ where $\mathbf{y}_S = (y_i)_{i \in S}$ and $\overline{\mathbf{y}}_S = \sum_{i \in S} y_i$. Let $\mathcal{F}_S$ denote the set of feasible allocations for coalition S.

Constrained or household core. Suppose an immutable household structure P and that household members or factions cannot entertain outside liaisons without the knowledge, approval or at least tacit agreement of the rest of the household. Therefore, if a household member belongs to a coalition, then implicitly the entire household is part of that coalition. Therefore, since the household structure P is cemented, the only coalitions of interest here are those composed of households in P. Let $\mathcal{G}(P)$ denote the family of non-empty subsets of P. For $G \in \mathcal{G}(P)$, define $C(G) := \bigcup_{h \in G} h$. $C(G)$ is the coalition consisting of all the constituents of all the households in G and $G \in \mathcal{Q}_{C(G)}$. The fixed household structure P implies that G is the only possible household structure for coalition $C(G)$. Finally, we define $\mathcal{X}(P) = \{\mathbf{x} \in \mathcal{X} : (\mathbf{x}, P) \in \mathcal{F}\}$, the set of feasible commodity allocations given the household structure P.

Now we are prepared to define $\mathcal{C}^+(P)$, the
strong P-core or
strong household core w.r.t. P or
strong constrained core w.r.t. P.

Formally, $\mathcal{C}^+(P) = \{\mathbf{x} \in \mathcal{X}(P) :$ There is no $S \in \mathcal{H}$ such that $(S, \mathbf{x}) \in \mathbb{I}_P.\}$ where $\mathbb{I}_P$ consists of the pairs $(S, \mathbf{x})$ with $\mathbf{x} \in \mathcal{X}(P)$ and $S = G(G)$ for some $G \in \mathcal{G}(P)$ together with $(G, \mathbf{y}_S) \in \mathcal{F}_S$ such that

(i) $U_i(\mathbf{y_h}; h) \geq U_i(\mathbf{x_h}; h)$ for all $h \in G$, $i \in h$,
(ii) $U_i(\mathbf{y_h}; h) > U_i(\mathbf{x_h}; h)$ for some $h \in G$, $i \in h$.

The corresponding "weak core" $\mathcal{C}^-(P)$ contains the strong core and is possibly larger. It obtains when (i) and (ii) are replaced by

(iii) $U_i(\mathbf{y_h}; h) > U_i(\mathbf{x_h}; h)$ for all $h \in G$, $i \in h$.

Existence of strong P-core allocations is not at issue if P-equilibria exist. This follows from the following P-core inclusion result.

Proposition 8.1 (Haller (2000), Proposition 3) *If* $(p; \mathbf{x})$ *is a* P*-equilibrium and (BE) holds, then* $\mathbf{x}$ *belongs to* $\mathcal{C}^+(P)$*, the strong* P*-core.*

Full core. The presumption here is that both commodities and consumers can be freely allocated. Still, we posit that if part of a household belongs to a coalition, then the entire household belongs to that coalition. But now we allow for an endogenous household structure so that dissenting household members have the option to secede, try to form new households and join coalitions including such newly formed households. Accordingly, the **strong (full) core** is defined as $\mathcal{C}^+ = \{(\mathbf{x}; P) \in \mathcal{F} :$ There is no $S \in \mathcal{H}$ such that $(S, (\mathbf{x}; P)) \in \mathbb{I}.\}$ where $\mathcal{F}$ denotes the set of feasible allocations of the economy and the improvement relation $\mathbb{I}$ consists of the pairs $(S, (\mathbf{x}; P)) \in \mathcal{H} \times \mathcal{F}$ for which there exists $(\mathbf{y}_S = (y_i)_{i \in S}; Q) \in \mathcal{F}_S$ such that

(i) $U_i(\mathbf{y}_{\mathbf{Q(i)}}; Q(i)) \geq U_i(\mathbf{x}_{\mathbf{P(i)}}; P(i))$ for all $i \in S$,
(ii) $U_i(\mathbf{y}_{\mathbf{Q(i)}}; Q(i)) > U_i(\mathbf{x}_{\mathbf{P(i)}}; P(i))$ for some $i \in S$.

The corresponding "weak core" $\mathcal{C}^-$ obtains if (i) and (ii) are replaced by

(iii) $U_i(\mathbf{y}_{\mathbf{Q(i)}}; Q(i)) > U_i(\mathbf{x}_{\mathbf{P(i)}}; P(i))$ for all $i \in S$.

The core can be empty for various reasons. In the context of pure group formation, two-sided matching models à la Gale and Shapley (1962) have a non-empty core. However, they also present a roommate problem without a stable matching and, hence, with an empty core. Alkan (1988) provides an example of non-existence of stable threesome matchings and, therefore, non-existence of core allocations. In the absence of consumption externalities, the core concept introduced here and the core concept for club models coincide. Gilles and Scotchmer (1997) have shown that the core may be empty if preferences are non-convex. In Gersbach and Haller (1999), we presented an example that the weak core can be empty even if preferences are convex. The specifications of the example are as follows.

There is a single commodity, that is $\ell = 1$. There are three consumers labelled $i = 1, 2, 3$, with generic consumption bundles x_i. For a potential household h, its endowment is $\omega_h = |h|$. In particular, each individual has endowment $\omega_{\{i\}} = 1$. Moreover, we assume a weak form of separability and GAN, i.e.,

$$U_i(\mathbf{x_h}; h) = u(x_i) + v\left(\sum_{j \in h \setminus i} x_j\right) \quad \text{for } i \in I, h \in \mathcal{H}_i.$$

Specifically, we assume

$$u(r) = r \text{ and } v(r) = r - \frac{4}{9}r^2 \text{ for } r \geq 0.$$

Hence *ex ante*, consumers are completely homogeneous with respect to endowments and preferences.

In general, core inclusion cannot be expected. Example 5 in Gersbach and Haller (2011) has a pair of competitive equilibria with free exit whose allocations are strictly Pareto ranked. Thus the inferior equilibrium allocation does not belong to the weak core. But weak core inclusion obtains for the most demanding class of equilibria which, however, rarely exist.

Proposition 8.2 (Gersbach and Haller (2005), Proposition 4) *Let $(p, \mathbf{x}; P)$ be a competitive equilibrium at which no group benefits from forming a new household. Then $(\mathbf{x}; P) \in \mathcal{C}^-$.*

Adding absence of consumption externalities and some other assumptions yields strong core inclusion. A consumer i exhibits absence of consumption externalities if there exists a function $V_i : X_i \times \mathcal{H}_i \to \mathbb{R}$ such that $U_i(\mathbf{x_h}; h) = V_i(x_i; h)$ for $h \in \mathcal{H}_i$, $\mathbf{x_h} = (x_j)_{j \in h} \in \mathcal{X}_h$. Given absence of consumption externalities with preferences represented by $V_i : X_i \times \mathcal{H}_i \to \mathbb{R}$ and $(x_i, h) \in X_i \times \mathcal{H}_i$, we say that consumer i has a group preference against (x_i, h) if there exists $g \in \mathcal{H}_i$ such that $V_i(x_i; h) < V_i(y_i; g)$ for all $y_i \in X_i$.

Proposition 8.3 (Gersbach and Haller (2010), Proposition 3) *Suppose IPR and absence of consumption externalities with utility representations $V_i(x_i, h)$ that satisfy continuity and strict monotonicity in $x_i \in X_i$. If $(p, \mathbf{x}; P)$ is a competitive equilibrium at which no group benefits from forming a new household and no individual i has a group preference against $(x_i, P(i))$, then $(\mathbf{x}; P) \in \mathcal{C}^+$.*

As observed in Gersbach and Haller (2005), if instead the budget exhaustion property and the redistribution property of Gersbach and Haller (2001) are added to the hypothesis of Proposition 8.2, then again weak core inclusion can be replaced by strong core inclusion:

Proposition 8.4 *Suppose that both the budget exhaustion property and the redistribution property hold. Let $(p, \mathbf{x}; P)$ be a competitive equilibrium at which no group benefits from forming a new household. Then $(\mathbf{x}; P) \in \mathcal{C}^+$.*

References

Alkan, A. (1988). Nonexistence of stable threesome matchings. *Mathematical Social Sciences*, *16*, 201–209.

Gale, D., & Shapley, L. (1962). College admissions and the stability of marriage. *American Mathematical Monthly*, *92*, 261–268.

Gersbach, H., & Haller, H. (1999). Allocation among multi-member households: Issues, cores and equilibria. In A. Alkan, C. D. Aliprantis, & N. C. Yannelis (Eds.), *Current trends in economics: Theory and applications*. Heidelberg: Springer.

Gersbach, H., & Haller, H. (2001). Collective decisions and competitive markets. *Review of Economic Studies*, *68*, 347–368.

Gersbach, H., & Haller, H. (2005). When inefficiency begets efficiency. *Economic Theory*, *25*, 105–121.

Gersbach, H., & Haller, H. (2010). Club theory and household formation. *Journal of Mathematical Economics*, *46*, 715–724.

Gersbach, H., & Haller, H. (2011). Bargaining cum voice. *Social Choice and Welfare*, *36*, 199–225.

Gilles, R. P., & Scotchmer, S. (1997). Decentralization in replicated club economies with multiple private goods. *Journal of Economic Theory*, *72*, 363–387.

Haller, H. (1990). Large random graphs in presudo-metric spaces. *Mathematical Social Sciences*, *20*, 147–164.

Haller, H. (2000). Household decisions and equilibrium efficiency. *International Economic Review*, *41*, 835–847.

Part III
Other Forms of Group Formation

Chapter 9
Clubs, Matching, etc.

Several strands of theoretical literatures in game theory, economics and other social sciences deal with the formation of social and economic groups, most notably theories of clubs, matching, assignment games, and hedonic coalitions.

9.1 Clubs

Since the seminal contributions of Tiebout (1956) about local jurisdictions and local public goods and Buchanan (1965) about clubs both subjects have been widely investigated. Club models also deal with an endogenous partition of the population into groups. Some models allow for the competitive market allocation of multiple private goods as well. See in particular Cole and Prescott (1997), Ellickson (1979), Ellickson et al. (1999, 2001), Gilles and Scotchmer (1997, 1998), Wooders (1988, 1989, 1997). There are a variety of descriptive features distinguishing between club models and our model. First and foremost, in traditional club theory, the benefit of a club to a member is determined by its membership profile and/or the provision of local public goods (club goods) or abstract club projects. Thus individual welfare typically assumes the form $U_i(x_i, \Upsilon; h)$ where x_i is i's private consumption bundle, Υ stands for a bundle of club goods (local public goods) and/or discrete club projects and h is the set of club members. In serving this purpose, the club incurs a resource cost which it tries to recoup through the collection of admission fees. Procurement of private goods remains an individual decision and is subject to an individual budget constraint $px_i + \pi_i \leq p\omega_{\{i\}}$ where p is the market price for private goods and π_i is the club admission fee paid by consumer i which of course may depend on the chosen club. In contrast, the members of a group or household in our model—and if applicable, in models of hedonic coalitions, matching and assignment games—face a common budget constraint $p * \mathbf{x_h} \leq p \cdot \omega_h$ and reach a collective decision regarding the consumption of private goods.

In Gersbach and Haller (2010), we clarify in a more systematic way the relationship between our general equilibrium models with multi-member groups or

H. Gersbach and H. Haller, *Groups and Markets*,
DOI 10.1007/978-3-319-60516-6_9

households and club models with multiple private goods. For this purpose, we use "household model" as generic term for the kind of model developed and analyzed in previous chapters and use "club model" as generic term for a sophisticated club model which also allows for endogenous group formation and competitive market allocation of (multiple) private goods. For the sake of direct comparison, we rule out multiple club memberships, local public goods and abstract club projects, features which could be incorporated in a refined household model. Then the distinguishing feature of the club model is that individuals shop for both club memberships and private consumption. This means that club memberships are priced via admission fees or valuations. Each person makes optimal choices, given her individual budget constraint. In equilibrium, prices are such that markets for memberships and markets for commodities clear. Formally, with individual property rights and absence of consumption externalities:

Definition A state $(p, \mathbf{x}; P)$ is a **valuation equilibrium** if $(\mathbf{x}; P)$ is a feasible allocation and there exist admission prices or valuations $V_i(h)$ for $i \in I,\ h \in \mathcal{H}_i$, such that:

1. $\sum_{i \in h} V_i(h) = 0$ for $h \in P$;
2. $\sum_{i \in h} V_i(h) \leq 0$ for $h \in \mathcal{H},\ h \notin P$;
3. $px_i + V_i(P(i)) = p\omega_i$ for $i \in I$;
4. If $i \in I, h \in \mathcal{H}_i, y_i \in \mathcal{X}_i$ with $U_i(y_i; h) > U_i(x_i; P(i))$, then $py_i + V_i(h) > p\omega_i$.

In the household model, actual households make collective consumption decisions for their members, subject to a household budget constraint. In equilibrium, nobody wants to exercise an outside option (like becoming single) at the prevailing market prices and commodity markets clear. See Chap. 7 for the definition of our pertinent equilibrium concepts: CEFE, CEFH and a competitive equilibrium where no group of consumers can benefit from forming a new household. In both models, the outcome is an allocation of commodities to individual consumers and a partition of the population into clubs or households. The two models can be considered **equivalent** if they yield the same equilibrium outcomes.

A priori, the only difference between the two models lies in the equilibrium concepts. For our current purpose, the main implication of the analysis in Gersbach and Haller (2010) is that the set of CEFE and the set of CEFH contain the set of valuation equilibria in the absence of consumption externalities—and with continuous and strictly monotone preferences. This follows from the finding that under those assumptions, the standard equilibrium concept for club models, valuation equilibrium, conforms to the requirements of the most stringent equilibrium concept for the household model, a competitive equilibrium where no group of consumers can benefit from forming a new household.

Conversely, a competitive equilibrium $(p, \mathbf{x}; P)$ where no group of consumers can benefit from forming a new household constitutes a valuation equilibrium under the weak additional assumption that no individual i has a group preference against $(x_i; P(i))$, that is there does not exist $g \in \mathcal{H}_i$ such that $U_i(x_i; P(i)) < U_i(y_i; g)$

for all $y_i \in X_i$. Hence in a sense, the two models are equivalent in the absence of consumption externalities. Moreover, under the assumptions of the equivalence result, the respective equilibrium outcomes belong to the strong core, which is a rather general property of valuation equilibria.

The equivalence breaks down in both directions in the presence of consumption externalities. In Gersbach and Haller (2010), we present an example in which a Pareto-optimal valuation equilibrium allocation does not exist. However, there exists a competitive equilibrium where no group of consumers can benefit from forming a new household and the equilibrium allocation constitutes a strong Pareto-optimum. The equivalence also breaks down in the opposite direction: In Gersbach and Haller (2001, pp. 361–362), we present an example of a valuation equilibrium that fails to be weakly Pareto-optimal and to be a competitive equilibrium among households. This example further demonstrates that valuation equilibrium allocations may no longer belong to the strong core—and need not even belong to the weak core—in the presence of consumption externalities. The example is not a special or knife-edge case. For instance, Cobb–Douglas preferences in the example can be replaced by homothetic and strictly monotone preferences.

9.2 Two-sided Matching

Theories of two-sided matching and especially of marriage tend to deal exclusively with group formation and to neglect the allocation of economic resources. Needless to say that preferences for partners may embody material aspects albeit not explicitly. Perhaps the first formal treatment of marriage and matching is found in a combinatorial lemma known as the "marriage theorem", also known as Hall's theorem. It was first stated and shown in its definitive form in Hall (1935) and Maak (1935). Weyl (1949) introduced the term "marriage theorem". Jacobs (1969, pp. 105–106) provides a proof and statement of the theorem. We state it in slightly different form: *Let M be a finite non-empty set, F be a non-empty (possibly infinite) set, and $D : M \twoheadrightarrow F$ be a correspondence. Define $D(N) = \bigcup_{i \in N} D(i)$ for $N \subseteq M$. Then there exists an injective selection d of D if and only if $|D(N)| \geq |N|$ for all $N \subseteq M$.* The standard interpretation takes M as set of men, F as set of women, and $D(i)$ the set of i's acquaintances. In an injective selection d of D, each man is matched with exactly one acquaintance and each woman is matched with at most one man. There are conceivable alternative interpretations of a set $D(i)$ as the set of women whom i considers acceptable partners or as the set of women who find i acceptable.

The marriage market has been a prominent application of the two-sided matching approach (see Roth and Sotomayor (1990)) where individuals rank potential partners of the opposite sex. Gale and Shapley (1962) have shown in their seminal paper that there always exists a stable matching for any marriage market. Many later contributions have demonstrated the robustness of this classic result.

Our general framework encompasses deterministic matching models, often with the added feature that (b) groups (households) make collective decisions regarding the individual consumption plans of their members and (c) there is competitive exchange across households.

The obvious question is whether the existence results of the matching framework carry over to this general framework. Specifically, we will deal with the existence of competitive equilibria with free household formation in the *marriage market*. In the two-sided matching context, the condition that no consumer benefits from joining another household in our definition of a competitive equilibrium with free household formation (CEFH) is weaker than the stability condition in the matching literature (see Roth and Sotomayor (1990)) which requires that a matching be not blocked by any individual or pair of agents forming a new match. Therefore, this literature promises to provide further existence results in our context. We proceed in two steps. First, we note that the existence results carry over from the matching literature to our framework when there is only one commodity. Second, we find that with several commodities and active trade in markets, non-existence can occur.

9.2.1 Existence in the Marriage Market

In the first step, we consider a simple marriage market as follows. We suppose $\ell = 1$ and that the population is divided into two non-empty, finite and disjoint sets, M and F: $M = \{m_1, \ldots, m_m\}$ is the set of men, and $F = \{f_1, \ldots, f_n\}$ is the set of women. We assume that each individual has some endowment, $\omega_i > 0$ and $\omega_j > 0$, respectively, and that individual property rights prevail. The *preferences of men* are given by

$$U_i(\mathbf{x_h}; h) = x_i \qquad \text{if } h = \{m_i\} \tag{9.1}$$

$$U_i(\mathbf{x_h}; h) = x_i + g_{ij} \quad \text{if } h = \{m_i, f_j\} \tag{9.2}$$

$$U_i(\mathbf{x_h}; h) = x_i - \bar{g} \quad \text{in all other cases} \tag{9.3}$$

We assume $\bar{g} > 0$ and $0 \le g_{ij} \le \omega_i$ for any potential couple $\{m_i, f_j\}$. The *preferences of women* are defined accordingly. We call such preferences **pure group externalities of the matching type**.

Such a marriage market where utility can be freely transferred within a household by an appropriate allocation of commodities and no trade through markets occurs, can be viewed as a generalized assignment game. We obtain:

Proposition 9.1 *Suppose $\ell = 1$ and pure group externalities of the matching type. Then a competitive equilibrium with free household formation exists.*

Proof Recall that in a competitive equilibrium with free household formation (CEFH) as defined in Chap. 7,

3. no consumer can benefit from exit;
4. no consumer can benefit from joining another household.

Because of the exit condition 3 and $\bar{g} > 0$, we only have to consider single person households or matches between a man and a woman as potential households in a CEFH. Since our free household formation condition 4 is weaker than the stability condition in the matching literature we can rely on the existence proofs for the generalizations and variations of the assignment model provided by Shapley and Shubik (1972) in their seminal article and further by Quinzii (1984), Gale (1984) and Alkan and Gale (1990); see also Roth and Sotomayor (1990).

Let us check the essential assumptions as they are formulated in Alkan and Gale (1990), for example. Let us hypothetically extend the domain of U_i to negative consumption—which will not occur in equilibrium. Then the range of the utility function is all of $\mathbb{R}$, since $U_i(x_i)$ is unbounded above and below. Moreover, for any couple, the corresponding Pareto-frontier in utility space is linear. Hence, we can apply Theorem 1 of Alkan and Gale (1990) which establishes existence of a core payoff and, consequently, of a CEFH. □

9.2.2 Non-Existence in the Marriage Market

In the second step, we investigate the stable matching problem when not only individuals are matched but also commodities are traded and collective household decisions are taken. Then we encounter a number of new challenges.

We have seen that the existence results of the matching literature are applicable in our framework provided that there is only one commodity. With several commodities, however, households may actively trade in the market. Consequently, what is feasible for a household depends on market prices. This price-dependence tends to undermine existence, even if households are restricted to singles and heterosexual marriages. Although our equilibrium conditions 3 and 4 are weaker than the standard stability condition for the marriage market, the existence result for the special case $\ell = 1$ does not carry over to the multiple goods case as the following example demonstrates. In the three-person example, externalities are confined to the one female individual whom we shall call Anita. She experiences a positive group externality and a negative consumption externality (with respect to the second good) when living with a partner. For any constant consumption of the partner, the consumption externality becomes less severe as Anita's consumption of the second good increases.

If everyone is single, the market clearing price of the second good happens to be low. Then at the going prices, Anita can afford enough own consumption of the second good so that the positive group externality dominates and she and another individual can both benefit from forming a two-person household. Now suppose Anita belongs to a two-person household with one single person remaining. Then the market is cleared at a high relative price of the second good. Therefore, as Anita can afford too little consumption of the second good, the negative consumption externality dominates, and she is better off alone. Hence for any given household structure, the market clearing prices are such that Anita can benefit from a change of household. The details are as follows.

Example 9.1 (Example 3 of Gersbach and Haller (2011)) Let $\ell = 2$ and $I = \{1, 2, 3\}$ where the first two individuals are male and $i = 3$ is Anita, the only female. The individual endowments are given by:

$$w_1 = (0, 1), w_2 = (0, 1), w_3 = (1, 1).$$

Preferences are represented by utility functions of the form $U_i(\mathbf{x_h}; h)$. Specifically,

$$\begin{aligned}
&U_1(\mathbf{x_h}; h) = \ln x_1^2, && \text{if } h = \{1\}, \{1, 3\};\\
&U_2(\mathbf{x_h}; h) = \ln x_2^2, && \text{if } h = \{2\}, \{2, 3\};\\
&U_3(\mathbf{x_h}; h) = \alpha \ln x_3^1 + (1-\alpha) \ln x_3^2, && \text{if } h = \{3\};\\
&U_3(\mathbf{x_h}; h) = \alpha \ln x_3^1 + (1-\alpha) \ln(\max\{0, x_3^2 - kx_i^2\}) + g, && \text{if } h = \{3, i\}, i = 1, 2;
\end{aligned}$$

where $0 < \alpha < 1$, x_i^j denotes the quantity of good j $(j = 1, 2)$ consumed by individual i, and we adhere to the convention $\ln 0 = -\infty$.

Living in a two-person household with partner $i = 1$ or partner $i = 2$ provides the third individual with a positive group externality $(g > 0)$. She suffers, however, from a negative consumption externality $(1 > k > 0)$. We further assume that living in a three-person household or in $h = \{1, 2\}$ creates enormous negative group externalities and will never be chosen. Hence our model is of the matching type where the only conceivable household structures consist of single-person and two-person households.

Commodity prices are normalized so that $p_1 = 1$. Consider first the household structure $P^\circ = \{\{1\}, \{2\}, \{3\}\}$. It is obvious that there exists a unique competitive equilibrium $(p^0, \mathbf{x^0})$ relative to P° given by:

$$\begin{aligned}
p^0 &= (1, p_2^0)\\
x_1^0 &= (0, 1)\\
x_2^0 &= (0, 1)\\
x_3^0 &= (1, 1)
\end{aligned}$$

To determine the market clearing price, we observe that the demand x_3^2 is given by

$$x_3^2 = (1-\alpha)(1 + p_2)/p_2.$$

Therefore market clearing, $x_3^2 = 1$, yields $p_2^0 = \frac{1-\alpha}{\alpha}$. At the going equilibrium prices $i = 3$ could propose to $i = 1$ to form the household $h = \{1, 3\}$ by offering $i = 3$ one unit of commodity 2. The remaining problem of individual 3 is

$$\begin{aligned}
&\max \left\{\alpha \ln x_3^1 + (1-\alpha) \ln(\max\{0, x_3^2 - k\}) + g\right\}\\
&\text{s.t. } x_3^1 + p_2^0 x_3^2 = 1 + p_2^0.
\end{aligned}$$

The solution is

$$\hat{x}_3^2 = (1-\alpha)(1+p_2^0)/p_2^0 + \alpha k = 1 + \alpha k, \tag{9.4}$$
$$\hat{x}_3^1 \quad = 1 + p_2^0 - p_2^0 x_3^2 \quad = 1 - (1-\alpha)k \tag{9.5}$$

which yields utility

$$U_3(\mathbf{x_h}; h) = \alpha \ln(1-(1-\alpha)k) + (1-\alpha)\ln(1-(1-\alpha)k) + g = \ln(1-(1-\alpha)k) + g.$$

Suppose that we choose parameters (k, g) such that

$$\ln(1-(1-\alpha)k) + g > 0.$$

Then $(p^0, \mathbf{x^0};\ \mathrm{P}^\circ)$ is not a competitive equilibrium with free household formation because $h = \{1, 3\}$ will be formed at equilibrium prices.

Consider next the household structure $P^* = \{\{1, 3\}, \{2\}\}$. Consider household $h = \{1, 3\}$. The maximal utility the third individual can achieve, subject to 1's outside options, is attained when individual $i = 1$ consumes one unit of the second commodity. The remaining problem of individual 3 is as in the case before. Therefore we obtain the demand for the second commodity as

$$x_3^2 = (1-\alpha)(1+p_2)/p_2 + \alpha k.$$

But to be in equilibrium now, markets must clear again. Hence $x_3^1 = 1, x_3^2 = 1$ which requires equilibrium prices $p_2^* = \frac{1-\alpha}{\alpha(1-k)}$. The utility of individual 3 is

$$U_3(\mathbf{x}_\mathbf{h}^*; h) = (1-\alpha)\ln(1-k) + g.$$

Since there exist values of α such that

$$\ln(1-(1-\alpha)k) > (1-\alpha)\ln(1-k),$$

e. g. $\alpha = \frac{1}{2}$, we can fix such an α and choose parameter constellations (k, g) such that

$$U_3(\mathbf{x_h}; h) > 0 > U_3(\mathbf{x}_\mathbf{h}^*; h).$$

Since individual 3 can always achieve utility $U_3 = 0$ by living as a one-person household and consuming her endowments, we conclude that under the suitably chosen parameter constellation, $(p^*, \mathbf{x}^*;\ P^*)$ is not a competitive equilibrium with free household formation: agent 3 prefers to be single at the going market prices. However, we have established before that agent 3 prefers to form a two-person household at the market prices which would obtain if everybody were single. Since individuals 1 and 2 are completely interchangeable, we conclude that no CEFH exists. □□

This example shows that active trade across households poses a challenge with regard to existence of stable outcomes not only for us, but also for the traditional matching literature. The hypotheses of the example and of Proposition 9.1 differ in two respects. First, there are several commodities in the example. Second, there are no longer pure group externalities of the matching type. This begs the question whether existence of a CEFH can be obtained, if there are several commodities, but pure group externalities of the matching type prevail. In the most general form of the latter case, the population is partitioned into men and women; preferences are represented by $U_i(\mathbf{x_h}; h) = U_i^c(x_i) + U_i^g(h)$ such that based on the group preferences given by U_i^g alone, individual i strictly prefers staying single or forming a two-person household with a member of the opposite sex ("marriage") to any other household. Under these circumstances, the following proposition holds.

Proposition 9.2 *Suppose the general case of pure group externalities of the matching type. If*

(i) $(p, \mathbf{x})$ is a competitive equilibrium of the pure exchange economy represented by $(U_i^c, w_{\{i\}})_{i \in I}$ and
(ii) P is a stable matching with respect to pure group preferences,

then the state $(p, \mathbf{x}; P)$ is a CEFE.

The proof of the proposition is straightforward. According to the classical result of Gale and Shapley (1962), condition (ii) can always be satisfied. Under standard assumptions on consumer characteristics, condition (i) holds as well and, consequently, a CEFE with stable matching exists.

Needless to say that a CEFE of the form suggested by Proposition 9.2 happens to be a CEFH, if the assumptions of Proposition 9.1 hold. Furthermore, such a CEFE turns out to be a CEFH whenever the equilibrium household structure is the unique optimal household structure based solely on group preferences. This follows from

Proposition 9.3 *Suppose pure group externalities, that is*
$U_i(\mathbf{x_h}; h) = U_i^c(x_i) + U_i^g(h)$ for $\mathbf{x_h} \in \mathcal{X}_h$, $h \in \mathcal{H}_i$, $i \in I$. If

(i) $(p, \mathbf{x})$ is a competitive equilibrium of the pure exchange economy represented by $(U_i^c, \omega_{\{i\}})_{i \in I}$, where all U_i^c, $i \in I$, satisfy local non-satiation, and
(ii) P is the unique optimal household structure based solely on group preferences represented by U_i^g, $i \in I$,

then the allocation $(\mathbf{x}; P)$ is fully Pareto-optimal and the state $(p, \mathbf{x}; P)$ is a CEFH.

Proof Suppose the state $(p, \mathbf{x}; P)$ satisfies (i) and (ii). For $i \in I$, let $P(i)$ denote the corresponding element of P, i.e., the household to which i belongs. We first claim that there do not exist any $i \in I$ and $h \in \mathcal{H}_i$ with $U_i^g(h) > U_i^g(P(i))$. For otherwise, there would exist an optimal household structure based solely on group preferences, P^* such that $U_i^g(P^*(i)) \geq U_i^g(h) > U_i^g(P(i))$ and, therefore, $P^* \neq P$, contradicting (ii). Moreover, we observe that $\mathbf{x}$ is a Pareto-optimal allocation of the pure exchange economy $(U_i^c, \omega_{\{i\}})_{i \in I}$ because consumers are locally non-satiated.

Consider any feasible allocation $(\mathbf{y}; P')$ and $i \in I$. Suppose $U_i(\mathbf{y}_{\mathbf{P}'(\mathbf{i})}; P'(i)) > U_i(\mathbf{x}_{\mathbf{P}(\mathbf{i})}; P(i))$. Then we claim that $U_j(\mathbf{y}_{\mathbf{P}'(\mathbf{j})}; P'(j)) < U_j(\mathbf{x}_{\mathbf{P}(\mathbf{j})}; P(j))$ for some $j \in I$. From above we have $U_i^g(P'(i)) \leq U_i^g(P(i))$. Hence $U_i^c(y_i) > U_i^c(x_i)$. This implies $U_j^c(y_j) < U_j^c(x_j)$ for some $j \in I$, since $\mathbf{x}$ is a Pareto-optimal consumption allocation of the pure exchange economy $(U_i^c, \omega_{\{i\}})_{i \in I}$. Further $U_j^g(P'(j)) \leq U_j^g(P(j))$. Hence the claim follows. This shows that the allocation $(\mathbf{x}; P)$ is fully Pareto-optimal.

Next we prove that the state $(p, \mathbf{x}; P)$ is a competitive equilibrium with free household formation. Suppose not. Hence, there exist a household $h \in P$ and an individual $i \in h$ such that either i is better off as a single or there exists a household $g \in P$ which i can join and where the utility of all members of the newly created household $g \cup \{i\}$ can be improved. We concentrate on the latter case. The case when individual i forms a one-person household is similar.

Let $\mathbf{y}_{\mathbf{g} \cup \{\mathbf{i}\}} \in B_{g \cup \{i\}}(p)$ be an allocation in the newly created household $g \cup \{i\}$ which makes everybody in this household better off. Since $U_j^g(g \cup \{i\}) \leq U_j^g(P(j))$, $U_j^c(y_j) > U_j^c(x_j)$ has to hold for each $j \in g \cup \{i\}$. But since $(p, \mathbf{x})$ is a competitive equilibrium of the pure exchange economy $(U_i^c, \omega_{\{i\}})_{i \in I}$, we have $py_j > p\omega_j$ for all $j \in g \cup \{i\}$. Therefore, $p * \mathbf{y}_{\mathbf{g} \cup \{\mathbf{i}\}} > p \cdot \omega_{g \cup \{i\}}$, contradicting $\mathbf{y}_{\mathbf{g} \cup \{\mathbf{i}\}} \in B_{g \cup \{i\}}(p)$. □

If, however, in the last proposition (ii) is replaced by the requirement that P be the unique stable matching with respect to pure group preferences, then $(p, \mathbf{x}; P)$ still happens to be a CEFE, but may no longer be a CEFH, and $(\mathbf{x}; P)$ may not be fully Pareto-optimal. To see this, it suffices to consider a population consisting of one male and one female, where the male has a slight preference (in terms of the utility difference) for staying single and the female has a strong preference for being married. Let the corresponding (absolute) utility differentials be ϵ for the male and Δ for the female. Then the stable matching with respect to pure group preferences requires both to remain single. Now suppose they have identical and strictly positive endowments and identical consumption preferences of the Cobb–Douglas type. Then the competitive equilibrium in (i) is a no trade equilibrium. If ϵ is sufficiently small and Δ is sufficiently large, they can both benefit from getting married and shifting some consumption from the female to the male—which shows our claim.

Additional examples of non-existence appear in the literature on hedonic coalitions and matching. Example 4 of Bogomolnaia and Jackson (2002), the example of Alkan (1988) and the roommate example of Gale and Shapley (1962) all constitute purely hedonic cases that differ from marriage models. Example 2 of Gersbach and Haller (2011) does not belong to the marriage category either. It shares features of matching and assignment games due to the presence of a consumption good and pure group externalities. Example 9.1 above is reminiscent of Example 3.3 in Drèze and Greenberg (1980), despite the fact that the latter is not a marriage model. Their common feature consists in the interaction of household formation and commodity allocation. The striking feature of Drèze and Greenberg's example is the absence of any consumption or group externalities. It is driven by household-specific (coalition-specific) endowments w_h with $w_h \neq \sum_{i \in h} w_{\{i\}}$ for some households h; in other words, by the presence of endowment externalities.

9.2.3 Discussion

Non-existence of a competitive equilibrium with certain properties renders the discussion of equilibrium household structures and equilibrium welfare obsolete. There are several possible responses to the non-existence problem.

First, the model might be misspecified. For instance, the modeling might be too parsimonious. While household stability cannot be achieved on purely economic grounds, given the two types of outside options depicted here, a full account of all the forces that stabilize—or destabilize—households might restore equilibrium. For example, household-specific human capital can serve as a bond among household members. Furthermore, the market for marriages may be more competitive than reflected in our equilibrium concept.

Second, one might suspect that price-taking is too restrictive. If only consumers could freely recontract without regard to market prices, then the economy would settle in an equilibrium state in the sense of Edgeworth, that is a core allocation. Indeed, the full core which allows for the reallocation of consumers and commodities, happens to be non-empty in the previous example. However, Gersbach and Haller (1999) contains a three-person example where gender does not matter and the full core turns out to be empty.

Third, non-existence of equilibrium may capture an important feature of reality. Let us recapitulate the essence of the previous example. Individuals may find it optimal to split at the going market prices in order to reduce negative consumption externalities. But at equilibrium prices of the changed household structure, individuals may find it optimal to form a two-person household in order to benefit from group externalities, because they can buy more of those goods which generate less consumption externalities. The marital status of the woman in the example affects her market opportunities and vice versa. Therefore, the woman may simply go through a sequence of marriage, divorce, marriage, divorce, etc., which constitutes an example of sequential monogamy, possibly with breaks. A dynamic approach suggests itself for future analysis.

Finally, non-existence may simply be a small number or integer problem that goes away when the population is large enough. For instance, non-existence in Example 2 of Gersbach and Haller (2011) disappears after suitable replication. Insofar, non-existence may be considered merely an artifact of the particular example. However, the problem is more intricate. Non-existence in Example 9.1 does not vanish under replication. For suitable parameter values, one cannot find a price vector at which females would be indifferent between being single and forming a two-person household and at which markets for commodities would clear. A sizeable (relative to the economy) group of consumers of the same type keeps moving simultaneously into or out of households depending on the prices at which commodity markets clear.

The non-existence example is complementary to the non-existence problems and examples that Gilles and Scotchmer (1997) have developed for replicated club economies. Their example 1 (revisited) shows that trade among private commodities in club economies can lead to non-existence of competitive equilibria for any finite number of replications. The reason is that no finite economy has efficient scale. In our case, the problem is not efficient scale as replications cannot improve per capita utility. Non-existence occurs because of consumption externalities (and their interaction with group externalities) whose impact varies with the prices of the private commodities. Moreover, in our example active trade across households can lead to non-existence in a two-sided matching model while the example in Gilles and Scotchmer (1997) is not of the matching type.

There are two ways how existence can be restored. First, sufficient dispersion of consumer characteristics will restore existence. If each agent is replaced not by identical clones, but by similar yet non-identical copies, then at certain prices, some of the females may wish to remain single while others may wish to stay in two-person households and, consequently, the household structure may end up to be stable. This certainly works in the previous example. This is not to say that existence would never become a problem if only consumer characteristics were well dispersed.

Second, a continuum model may also restore existence. In Gersbach et al. (2015), we elaborate on the continuum version of Example 9.1. There is a continuum of Anitas of unit Lebesgue measure and a continuum of males of Lebesgue measure two. In equilibrium, a fraction ξ of Anitas will be matched. Incidentally, $\alpha = k = 1/2$, $g = \ln 2$ yields a unique $\xi = 2(\sqrt{5.5} - 2)$ which is an irrational number. Therefore, in that case there does not exist an equilibrium for any replica of the finite model—which confirms our earlier claim.

9.3 Other Models of Group Formation

In addition to matching models (as in Gale and Shapley (1962); Alkan (1988); Roth and Sotomayor (1990)), our general framework includes a variety of models of coalition formation, among others assignment games (e.g. Shapley and Shubik (1972); Roth and Sotomayor (1990)) and models with hedonic coalitions (e.g. Greenberg (1978); Bogomolnaia and Jackson (2002); Banerjee et al. (2001)). In the cited literature, markets are inactive and relative prices are irrelevant, simply because there exits at most one tradeable commodity. A noteworthy exception are Drèze and Greenberg (1980) who combine the concepts of individual stability and price equilibrium, but confine the analysis of their most comprehensive model to an instructive example as discussed in Sect. 9.2.2.

Recall that the concept of competitive equilibrium with free household formation (CEFH) requires 1. collective rationality of group decisions; 2. market clearing; 3. that no individual can benefit from exit, i.e., no individual wants to leave a group and participate as a one-member group in the market at the going equilibrium prices; 4. that no individual can benefit from joining another group, i.e., no individual can leave a group and can propose a feasible consumption allocation to the members of a new group, created by the individual and another already existing group, which makes everybody in the new group better off at the going equilibrium prices. Condition 4 is akin to individual stability in the sense of Drèze and Greenberg (1980). The conjunction of conditions 3 and 4 constitutes the weak version of individual stability in the sense of Bogomolnaia and Jackson (2002). Next we turn to similarities in specific restrictions on preferences. In their general model, Drèze and Greenberg (1980) assume absence of consumption externalities, that is utilities of the form $U_i(x_i; h)$ and refer to the dependence on h as the "hedonic aspect" of i's preferences. In their Example 3.3 alluded to earlier, both consumption and group externalities are absent. Bogomolnaia and Jackson (2002) among others call i's preferences "purely hedonic" if U_i depends only on h. We use the terminology "group externality" in lieu of "hedonic aspect".

References

Alkan, A. (1988). Nonexistence of stable threesome matchings: Note. *Mathematical Social Sciences, 16*, 201–209.

Alkan, A., & Gale, D. (1990). The core of the matching game. *Games and Economic Behavior, 2*, 203–212.

Banerjee, S., Konishi, H., & Sönmez, T. (2001). Core in a simple coalition formation game. *Social Choice and Welfare, 18*, 135–153.

Bogomolnaia, A., & Jackson, M. O. (2002). The stability of hedonic coalition structures. *Games and Economic Behavior, 38*, 201–230.

Buchanan, J. M. (1965). An economic theory of clubs. *Economica, New Series, 32*(125), 1–14.

Cole, H. L., & Prescott, E. C. (1997). Valuation equilibrium with clubs. *Journal of Economic Theory, 74*, 19–39.

Drèze, J., & Greenberg, J. (1980). Hedonic coalitions: Optimality and stability. *Econometrica, 48*, 987–1003.

Ellickson, B. (1979). Competitive equilibrium with local public goods. *Journal of Economic Theory, 21*, 46–61.

Ellickson, B., Grodal, B., Scotchmer, S., & Zame, W. R. (1999). Clubs and the market. *Econometrica, 67*, 1185–1218.

Ellickson, B., Grodal, B., Scotchmer, S., & Zame, W. R. (2001). Clubs and the market: Large finite economies. *Journal of Economic Theory, 101*, 40–77.

Gale, D. (1984). Equilibrium in a discrete exchange economy with money. *International Journal of Game Theory, 13*, 61–64.

Gale, D., & Shapley, L. (1962). College admissions and the stability of marriage. *American Mathematical Monthly, 92*, 261–268.

Gersbach, H., Haller, H., & Konishi, H. (2015). Household formation and markets. *Economic Theory, 59*, 461–507.

Gersbach, H., & Haller, H. (1999). Allocation among multi-member households: Issues, cores and equilibria. In A. Alkan, C. D. Aliprantis, & N. C. Yannelis (Eds.), *Current trends in economics: Theory and applications*. Heidelberg: Springer.

Gersbach, H., & Haller, H. (2001). Collective decisions and competitive markets. *Review of Economic Studies, 68*, 347–368.

Gersbach, H., & Haller, H. (2010). Club theory and household formation. *Journal of Mathematical Economics, 46*, 715–724.

Gersbach, H., & Haller, H. (2011). Competitive markets, collective decisions and, group formation. *Journal of Economic Theory, 146*, 275–299.

Gilles, R. P., & Scotchmer, S. (1998). Decentralization in club economies: How multiple private goods matter. In D. Pines, E. Sadka, I. Z., (Eds.), *Topics in public economics: Theoretical and applied analysis*, Chap. 5. Cambridge, UK: Cambridge University Press.

Gilles, R. P., & Scotchmer, S. (1997). Decentralization in replicated club economies with multiple private goods. *Journal of Economic Theory, 72*, 363–387.
Greenberg, J. (1978). Pure and local public goods: A game-theoretic approach. In A. Sandmo (Ed.), *Essays in public economics*. Lexington, MA: Heath and Co.
Hall, P. (1935). On representatives of subsets. *Journal of the London Mathematical Society, 10*, 26–30.
Jacobs, K. (1969). *Selecta mathematica I*. HTB 49. Heidelberg: Springer.
Maak, W. (1935). Eine neue definition der fastperiodischen funktionen. *Abhandlungen aus dem Mathematischen Seminar Hamburg, 11*, 240–244.
Quinzii, M. (1984). Core and competitive equilibria with indivisibilities. *International Journal of Game Theory, 13*, 41–60.
Roth, A. E., & Sotomayor, M. A. O. (1990). *Two-sided matching: A study in game-theoretic modeling and analysis*. Cambridge, UK: Cambridge University Press.
Shapley, L. S., & Shubik, M. (1972). The assignment game I: The core. *International Journal of Game Theory, 1*, 111–130.
Tiebout, C. M. (1956). A pure theory of local expenditures. *Journal of Political Economy, 64*, 416–424.
Weyl, H. (1949). Almost periodic invariant vector sets in a metric vector space. *American Journal of Mathematics, 71*, 178–205.
Wooders, M. H. (1988). Stability of jurisdiction structures in economies with local public goods. *Mathematical Social Sciences, 15*, 29–49.
Wooders, M. H. (1989). A Tiebout theorem. *Mathematical Social Sciences, 18*, 33–55.
Wooders, M. H. (1997). Equivalence of Lindahl equilibrium with participation prices and the core. *Economic Theory, 9*(2), 115–127.

Chapter 10
Related Work

10.1 Related Literature

Our tour d'horizon of the literature is far from comprehensive and cannot do justice to all important contributions. The theoretical work most closely related to ours deals with endogenous group formation. The most prominent strands are the club literature on the one hand and the matching literature on the other hand. The empirical literature is mainly devoted to household decision making. A detailed and lucid exposition of the main issues and the early literature is offered by Bergstrom (1997). Apps and Rees (2009) provide an elaborate description and discussion of various theories of household decisions and the connection between models of two-person households and public economics.

The "marriage theorem" of Hall (1935) and Maak (1935) is the first mathematical result related to marriage and matching we are aware of. The two-sided matching approach pioneered by Gale and Shapley (1962) gave rise to a sizeable literature; see the references in Chap. 9. Gale and Shapley also examine an example of a roommate or one-sided matching problem. Formally, models of roommate markets (one-sided matching) encompass "marriage markets" and two-sided matching as a special case. Study of these markets remains a very active research area to this day. See, e.g., Klaus (2011) and Sönmez and Pathak (2013). Alternative models have focused on different aspects of group formation and provided additional insights, like models of clubs (elaborated on in Chap. 9), hedonic coalitions (e.g., Greenberg (1978), Bogomolnaia and Jackson (2002), Banerjee et al. (2001)), assignment games (see for instance, Shapley and Shubik (1972), Roth and Sotomayor (1990)), and multilateral bargaining (for example Bennett (1988, 1997), Crawford and Rochford (1986), Rochford (1984)).

The seminal contribution of Becker (1973) has included social and economic traits of households and families in a model of mating and matching. Becker (1965, 1971, 1973, 1978, 1981) has demonstrated that an inquiry into the determinants of and

H. Gersbach and H. Haller, *Groups and Markets*,
DOI 10.1007/978-3-319-60516-6_10

connections between sociological and economic choices can be very productive.[1] Our research agenda is in the spirit of Becker. But we use a different model and address different questions. For instance, household-specific externalities play an important role in our approach. In contrast, Becker's model avoids consumption externalities in a unique way, by introducing a "household good", the sole explicit consumption good which is non-tradeable, yet perfectly divisible within each household and does not cause any consumption externalities. Most importantly, our investigation of interacting allocation and decision mechanisms incorporates the collective rationality concept of Chiappori (1988b, 1992) into a general equilibrium framework. That concept has been extensively examined and applied in the empirical literature. See the critical assessment, also from an empirical perspective, by Apps and Rees (2009). We adopt and explore the concept for purely theoretical analysis:

> Each household makes an efficient collective decision, i.e., given prices and wages, the household takes a decision regarding individual consumption of its members which is Pareto efficient within the household. In other words, the household chooses an element in its "efficient budget set".

In general, this is the only assumption on household behavior we make if the household structure is taken as exogenously given. In case households are endogenously formed, there is, as a rule, a surplus to be divided by household members, for instance a surplus relative to the outside opportunities they would have as single individuals. The creation and sharing of such a surplus constitutes the rationale for household formation in the first place. But how will any potential surplus be divided among household members? Browning et al. (2006, p. 6) list a number of different approaches to model intra-household bargaining. They further state—and we concur—that there is no broad consensus which particular model to use. We follow them and many others and assume collective rationality of households.

Traditionally, economic theory and empirical research have treated households as if they were single consumers, despite the fact that many households consist of several members. This is commonly known as the *unitary model* of the household. The underlying assumption is that there exists a representative consumer for the household. As a practical matter, household expenditure data, for instance U.S. Census data commonly used in empirical research may report the composition of households without disaggregating household consumption (expenditure) and factor supply (income) with respect to household members. Still, one would want to perform a more detailed description and analysis of household composition and behavior on both empirical and theoretical grounds. In fact, the unitary model has been rejected repeatedly (Blundell et al. (2005)). The collective rationality model has become the main alternative—but by no means the only alternative. It has been intensively investigated in the empirical literature and some related theoretical work. For instance,

[1]The equilibrium analysis of Grossbard-Shechtman (1984) is motivated by Becker's approach, views marriage and labor markets as mutually related and models marriage as an exchange of household labor between spouses.

Fortin and Lacroix (1997) find efficiency in labor supply and Browning and Chiappori (1998), Donni (2009) find efficient household commodity demand. Parametric test results by Browning et al. (1994) and Horney and McElroy (1988) suggest that the collective rationality model performs better than the unitary (representative consumer) model. See also the surveys by Bourguignon and Chiappori (1992, 1994) and Kapteyn and Kooreman (1992). Snyder (1999) has proposed nonparametric tests based on revealed preference theory and quantifier elimination techniques. Snyder (2000) finds for a sample from the National Longitudinal Surveys that all 265 households satisfy the restrictions of the collective rationality model, whereas 259 thereof satisfy the restrictions of the representative consumer model. For further revealed preference tests, see Cherchye et al. (2013) and references therein.

Given that one assumes collective rationality, a particular efficient household decision can always be obtained as the outcome of maximizing a utilitarian welfare function subject to the budget constraint (and non-negativity constraints). Even if the welfare or Pareto weights are taken as exogenous by the household, they are not necessarily exogenous or constant over time. The literature (Browning et al. (2013); Browning et al. (2006); Chiappori and Ekeland (2006); e.g.) tends to distinguish between so-called distribution factors and prices as variables that influence intra-household balance of bargaining power. We follow Basu (2006) and distinguish between endogenous, denoted x, and exogenous, denoted z, determinants of intra-household bargaining power. Labor incomes would constitute endogenous factors and might be part of x. Exogenous factors could be non-labor income(s), legal provisions, the sex ratio in the marriage market, individual wealth at the time of household formation, etc. In the context of Basu (2006), $x = x(\alpha)$ and $\alpha = \widehat{\alpha}(x, z)$ where α is the weight of the "first member" of a couple. Basu considers two conceivable scenarios: First, given z, the values of α and x are endogenously and simultaneously determined where α is a fixed point of the composed mapping $\alpha \mapsto \widehat{\alpha}(x(\alpha), z)$ and $x = x(\alpha)$. Second, still $x = x(\alpha)$, but α adjusts to endogenous factors with a time lag: x_t, z_t, and α_t follow a dynamic process in discrete time t where $x_t = x(\alpha_t)$, $\alpha_t = \widehat{\alpha}(x_{t-1}, z_t)$. The latter scenario is the more plausible one.

Another special case of efficient household decision making is Nash bargaining. Manser and Brown (1980), McElroy and Horney (1981), Horney and McElroy (1988), and McElroy (1990) have developed parametric models to discern testable properties of household consumption plans resulting from cooperative bargaining. See also the critique by Chiappori (1988a), the reply by McElroy and Horney (1990), and the rejoinder by Chiappori (1991). The assumption of Nash-bargained household consumption allows for comparative statics with respect to the distribution of bargaining power in some or all households, like in Gersbach and Haller (2009). Mutatis mutandis, our above comments regarding welfare weights also apply to bargaining weights.

The collective rationality model is not the only conceivable alternative to the unitary model as Browning et al. (2006) report. Several important contributions argue against assuming efficiency in household decision making (Lundberg and Pollak (2003), Konrad and Lommerud (1995, 2000)). For instance, Lundberg et al. (1997) provide evidence that tends to support the idea that household members do not pool their incomes as it would be implied by efficient collective decision making. Further evidence against collective rationality of households is presented by, among others, Fortin and Lacroix (1997) who reject efficiency in labor supply for couples with young children, Udry (1996) who rejects efficiency in household agricultural production, and Dercon and Krishnan (2000) and Robinson (2012) who find that families do not efficiently share risk. Lundberg and Pollak (1994) consider non-cooperative bargaining models. Lechene and Preston (2011) contrast the demand properties for non-cooperative households with those for collective models. Gersbach and Haller (2005, 2006) examine the nature and sustainability of inefficient household decisions in the general equilibrium context.

References

Apps, P. F., & Rees, R. (2009). *Public economics and the household*. Cambridge, UK: Cambridge University Press.

Banerjee, S., Konishi, H., & Sönmez, T. (2001). Core in a simple coalition formation game. *Social Choice and Welfare*, *18*, 135–153.

Basu, K. (2006). Gender and say: A model of household behaviour with endogenously determined balance of power. *The Economic Journal*, *116*, 558–580.

Becker, G. S. (1971). *Human capital: A theoretical and empirical analysis with special reference to education* (2nd ed.). New York: Columbia University Press.

Becker, G. S. (1973). A theory of marriage, part I. *Journal of Political Economy*, 81, 813–846. Reproduced as Chap. 11. In R. Febrero & P. S. Schwartz (Eds.), *The Essence of Becker* (1995). Stanford, CA: Hoover Institution Press.

Becker, G. S. (1965). A theory of the allocation of time. *Economic Journal*, *75*, 493–517.

Becker, G. S. (1978). *The economic approach to human behavior*. Chicago, IL: The University of Chicago Press.

Becker, G. S. (1981). *A treatise on the family*. Cambridge, MA: Harvard University Press.

Bennett, E. (1988). Consistent bargaining conjectures in marriage and matching. *Journal of Economic Theory*, *45*, 392–407.

Bennett, E. (1997). Multilateral bargaining problem. *Games and Economic Behavior*, *19*, 151–179.

Bergstrom, T. C. (1997). A survey of theories of the family. In M. Rosenzweig & O. Stark (Eds.), *Handbook of population and family economics, Chap. 2* (pp. 21–79). Amsterdam: Elsevier.

Blundell, R., Chiappori, P.-A., & Meghir, C. (2005). Collective labor supply with children. *Journal of Political Economy*, *113*, 1277–1306.

Bogomolnaia, A., & Jackson, M. O. (2002). The stability of hedonic coalition structures. *Games and Economic Behavior*, *38*, 201–230.

Bourguignon, F., & Chiappori, P.-A. (1992). Collective models of household behavior. *European Economic Review*, *36*(3–6), 355–364.

Bourguignon, F., & Chiappori, P.-A. (1994). The collective approach to household behavior. In R. Blundell, I. Preston, & I. Walker (Eds.), *The measurement of household welfare* (pp. 70–85). Cambridge, UK: Cambridge University Press.

Browning, M., Chiappori, P.-A., & Lewbel, A. (2013). Estimating consumption economies of scale, adult equivalence scales, and household bargaining power. *Review of Economic Studies*, *80*, 1267–1303.

Browning, M., Bourguignon, F., Chiappori, P.-A., & Lechene, V. (1994). Incomes and outcomes: A structural model of intrahousehold allocation. *Journal of Political Economy*, *20*, 1067–1096.

Browning, M., & Chiappori, P.-A. (1998). Efficient intra-household allocations: A general characterisation and empirical tests. *Econometrica*, *66*, 1241–1278.

Browning, M., Chiappori, P.-A., & Lechene, V. (2006). Collective and unitary models: A clarification. *Review of Economics of the Household*, *4*, 5–14.

Cherchye, L., De Rock, B., & Platino, V. (2013). Private versus public consumption within groups: Testing the nature of goods from aggregate data. *Economic Theory*, *54*, 485–500.

Chiappori, P.-A. (1988a). Nash-bargained household decisions: A comment. *International Economic Review*, *29*, 791–796.

Chiappori, P.-A. (1988b). Rational household labor supply. *Econometrica*, *56*, 63–89.

Chiappori, P.-A. (1991). Nash-bargained household decisions: A rejoinder. *International Economic Review*, *32*, 761–762.

Chiappori, P.-A. (1992). Collective labor supply and welfare. *Journal of Political Economy*, *100*, 437–467.

Chiappori, P.-A., & Ekeland, I. (2006). The micro economics of group behavior: General characterization. *Journal of Economic Theory*, *130*, 1–26.

Crawford, V. P., & Rochford, S. C. (1986). Bargaining and competition in matching markets. *International Economic Review*, *27*, 329–348.

Dercon, S., & Krishnan, P. (2000). In sickness and in health: Risk sharing within households in rural ethiopia. *Journal of Political Economy*, *108*, 688–727.

Donni, O. (2009). A simple approach to investigate intrahousehold allocation of private and public goods. *The Review of Economics and Statistics*, *91*, 617–628.

Fortin, B., & Lacroix, G. (1997). A test of the unitary and collective models of household labour supply. *Economic Journal*, *107*, 933–995.

Gale, D., & Shapley, L. (1962). College admissions and the stability of marriage. *American Mathematical Monthly*, *92*, 261–268.

Gersbach, H., & Haller, H. (2005). When inefficiency begets efficiency. *Economic Theory*, *25*, 105–121.

Gersbach, H., & Haller, H. (2006). Household inefficiency and equilibrium efficiency. In C. Schultz & K. Vind (Eds.), *Institutions, equilibria and efficiency: Essays in honor of Birgit Grodal* (pp. 187–209). Heidelberg: Springer.

Gersbach, H., & Haller, H. (2009). Bargaining power and equilibrium consumption. *Social Choice and Welfare*, *33*, 665–690.

Greenberg, J. (1978). Pure and local public goods: A game-theoretic approach. In A. Sandmo (Ed.), *Essays in public economics*. Heath and Co: Lexington, MA.

Grossbard-Shechtman, A. (1984). A theory of allocation of time in markets for labour and marriage. *The Economic Journal*, *94*, 863–882.

Hall, P. (1935). On representatives of subsets. *Journal of the London Mathematical Society*, *10*, 26–30.

Horney, M. J., & McElroy, M. B. (1988). The household allocation problem: Empirical results from a bargaining model. *Research in Population Economics*, *6*, 15–38.

Kapteyn, A., & Kooreman, P. (1992). Household labor supply: What kind of data can tell us how many decision makers there are? *European Economic Review*, *36*, 365–371.

Klaus, B. (2011). Competition and resource sensitivity in marriage and roommates markets. *Games and Economic Behavior*, *72*, 172–186.

Konrad, K., & Lommerud, K. E. (1995). Family policy with non-cooperative families. *Scandinavian Journal of Economics*, *97*, 581–601.

Konrad, K., & Lommerud, K. E. (2000). The bargaining family revisited. *Canadian Journal of Economics*, *33*, 471–487.

Lechene, V., & Preston, I. (2011). Noncooperative household demand. *Journal of Economic Theory, 146*, 504–527.
Lundberg, S., & Pollak, R. A. (1994). Non-cooperative bargaining models of marriage. *American Economic Review, Papers and Proceedings, 84*, 132–137.
Lundberg, S., & Pollak, R. (2003). Efficiency in marriage. *Review of Economics of the Household, 1*, 153–167.
Lundberg, S., Pollak, R., & Wales, T. (1997). Do husband and wives pool their resources? Evidence from the U.K. child benefit. *Journal of Human Resources, 32*, 463–480.
Maak, W. (1935). Eine neue Definition der fastperiodischen Funktionen. *Abhandlungen aus dem Mathematischen Seminar der Universität Hamburg, 11*, 240–244.
Manser, M., & Brown, M. (1980). Marriage and household decision-making: A bargaining analysis. *International Economic Review, 21*, 31–44.
McElroy, M. B. (1990). The empirical implications of Nash-bargained household behavior. *Journal of Human Resources, 25*, 559–583.
McElroy, M. B., & Horney, M. J. (1981). Nash-bargained household decisions: Toward a generalization of the theory of demand. *International Economic Review, 22*, 333–350.
McElroy, M. B., & Horney, M. J. (1990). Nash-bargained household decisions: Reply. *International Economic Review, 31*, 237–242.
Robinson, J. (2012). Limited insurance within the household: Evidence from a field experiment in kenya. *American Economic Journal: Applied Economics, 4*, 140–164.
Rochford, S. C. (1984). Symmetrically pairwise-bargained allocations in an assignment market. *Journal of Economic Theory, 34*, 262–281.
Roth, A. E., & Sotomayor, M. A. O. (1990). *Two-sided matching: A study in game-theoretic modeling and analysis*. Cambridge, UK: Cambridge University Press.
Shapley, L. S., & Shubik, M. (1972). The assignment game I: The core. *International Journal of Game Theory, 1*, 111–130.
Snyder, S. K. (1999). Testable restrictions of pareto optimal public good provision. *Journal of Public Economics, 71*, 97–119.
Snyder, S. K. (2000). Nonparametric testable restrictions of household behavior. *Southern Economic Journal, 67*, 171–185.
Sönmez, T., & Pathak, P. A. (2013). School admissions reform in Chicago and England: Comparing mechanisms by their vulnerability to manipulation. *American Economic Review, 103*, 80–106.
Udry, C. (1996). Gender, agricultural production, and the theory of the household. *Journal of Political Economy, 104*, 1010–1046.

Part IV
Extensions and Applications

Chapter 11
Power in General Equilibrium

In this chapter we explore power—how it can be integrated into our framework, how it affects individuals, and whether some level of power is compatible with a welfare perspective, or even desirable to enhance welfare.

11.1 The Notion of Power

First of all, it is important to clarify the notion of power. "Power" can have very different meanings in economics—and even more so in social science. Concepts such as market power, purchasing power, veto power, agenda-setting power, voting power, bargaining power, and power indices are commonly used. Needless to say, numerous notions of power have been forwarded by other disciplines, e.g., political science, sociology and philosophy. At a very general level, "power" refers to the possibilities an individual (or firm) has in a particular context, possibilities that others do not have or only have to a lesser extent. There are also situations in which all individuals have the same power (e.g., every citizen's voting power in a democracy). Finally, while there can be an "absence of power" for all individuals in a concept such as market power, some individuals *must* have some form of power in other contexts, such as agenda-setting power, for example, unless those decisions are left to an expert system or random device.

A central conception of power is an individual's ability to influence the decisions taken by the group to which he/she belongs. This kind of power may stem, among other factors, from certain formal rights and cause the denial of others. For instance, a top manager may not be allowed to join a labor union that conducts collective bargaining with the manager's company. In other contexts, the ability to impact a decision taken by a group may stem from individual effort or personal characteristics such as competence or charisma.

H. Gersbach and H. Haller, *Groups and Markets*,
DOI 10.1007/978-3-319-60516-6_11

In our framework, we will adhere to the conception of power connected to the ability to affect household decisions, and consider power in groups. Let us start with the following natural distinction:

- **Formal power**: This is the individual's say in the group's decisions, captured, e.g., by this individual's weight in a household welfare function or by his/her relative bargaining power in Nash-bargained household decisions.
- **Real power**: This is the additional utility an individual can achieve in a household, compared to his or her stand-alone utility as a single person.

We will proceed in three steps, based on five of our papers that address power: First, we study changes in formal power (Gersbach and Haller (2009a, 2012)); second, we provide two ways how formal power can be endogenized (Gersbach and Haller (2009b, 2011)); third, we study the relationship between formal and real power (Gersbach and Haller (2008)).

11.2 Changes in Formal Power

In a first step, we will investigate how changes in formal power affect the utility of individuals in the presence of a given household structure, i.e., the utility of those with more formal power after the change and of those with less. Societies often experience a shift of bargaining power in households. For instance, *ceteris paribus*, a shift of bargaining power in favor of the female partner has taken place over the last decades.

It is the consequences, not the causes of shifts in intra-household bargaining power that we examine in this chapter. We are concerned with pure economic (positive) effects on the allocation of resources, as well as welfare (normative) effects at both the individual and societal levels. As can be expected in a general equilibrium setting, these effects are subtle. They depend on whether or not equilibrium prices are affected by variations of some individuals' power and on the direction and magnitude of the price response. We study change of formal power in pure exchange economies and in the presence of production.

First, in Sect. 11.5, we study in detail how a shift of bargaining power within two-person households operating in competitive markets for consumption goods affects equilibrium allocation and welfare. We illustrate that an individual gains from an increase of bargaining power, necessarily to the detriment of others if price effects are sufficiently small. If price changes are drastic, the welfare of all household members moves in the same (positive or negative) direction when bargaining power shifts, at the expense (or for the benefit) of outside consumers. Sophisticated households might succeed in an attempt to manipulate the market outcome, not by misrepresenting endowments or individual preferences, but by misrepresenting the internal bargaining power.

Typically a shift of bargaining power within a set of households also affects other households. The most striking results occur when households are composed of two members of two different sociological groups, where a sociological group may be characterized by similar preferences or externalities. We find that each individual of a sociological group tends to benefit if he can increase his bargaining power, but suffers (or benefits less) if others in his group do the same. For quasi-linear preferences, however, a change of the bargaining power within a particular household only impacts on the distribution of the numéraire good within the household under consideration without affecting the consumption of other commodities, so that other households are not impacted at all.

Second, in Gersbach and Haller (2012) we examine changes in bargaining power within households in the presence of production and an active labor market. In particular, we study how a shift of bargaining power in favor of one partner in a household affects consumption and labor supply of both partners. We further investigate how the shift of bargaining power in a particular household causes a change of market prices and impacts upon other households. The case where such shifts occur in many households is considered as well. The most striking effects occur when household members differ in individual preferences and enjoy positive leisure-dependent externalities.

We embed a simple labor supply model in a general equilibrium model and distinguish between two goods: labor (leisure) and a composite consumption good. Individuals are endowed with time, part of which they enjoy as leisure and part of which they supply to the labor market. Labor is demanded by competitive firms that produce the composite consumption good.

We distinguish between a working class and a leisure class. The working class members receive only labor income and form the two-person households under consideration. The leisure class members receive all capital income and no labor income. We further allow that members in a household differ with regard to the weight they put on time they consume as leisure. Accordingly, we distinguish between hard workers and hedonists. Typically, a household consists of a hard worker and a hedonist.

Our first central result is that hard workers are bad company. If the hard worker in a household gains more influence, the partner is induced to work more and consume less, whereas the hard worker tends to work less and consume more. Moreover, through the labor market and the associated changing of the wages, the hard worker causes negative spillovers to the members of other households, both hard workers and hedonists. The most striking result is that a global change in relative bargaining power, where the hardworking member becomes more influential in each working class household, can render the entire working class worse off—with the leisure class as the sole beneficiary. We note that a similar effect cannot happen in a model without production or with the production sector fully owned by the working class.

Our second important finding is that a binding restriction on the number of hours an individual is allowed to work can benefit all workers, those for whom the restriction is binding and those for whom it is not. This finding represents a new argument why limits on working hours might be chosen in countries such as the Nordic countries.

At a more general level, the analysis shows that spillovers between households are not necessarily mitigated by the presence of a production sector. Moreover, new feedback effects occur in the presence of a production sector: A change in a household's labor supply alters aggregate labor supply and real wage rates which in turn affect all households.

11.3 Endogenizing Power

In a second step, we will study how formal power can be endogenized. We explore two ways how formal power, in its manifestation of weights in a household welfare function or in bargaining power, can be traced back to other characteristics such as abilities and preferences of human beings when household structures are not necessarily fixed and alternative social arrangements are considered.

11.3.1 Power of Voice

The power of voice presumes a framework with endogenous group formation. The underlying idea developed in Gersbach and Haller (2011) is as follows: Members of a group use outside options in two different ways. First, the possibility to exit and/or to join other existing groups determines the actual outside options—which constitute the reservation utilities of group members. Second, hypothetical outside options impact on the relative bargaining power in existing groups. Namely, individuals also reflect on hypothetical groups, whose formation would require that several individuals break away from their current group to form a new one. This (hypothetical) opportunity is articulated in the bargaining process—although ultimately, these new groups will not be formed. The best possible hypothetical scenarios for each person determine the person's relative bargaining power in her current group. This impact of articulating one's conceivable opportunities in hypothetical groups is called the "power of voice".

At a more abstract level, the power of voice suggests a way to formalize how discussion among individuals can bring about a consensus. In such a discussion, each side outlines the feasibility of potential best alternatives. Each individual assesses the feasibility of hypothetical outside options of other group members. The power of voice presumes that such an exchange transforms the best hypothetical outside alternative of one individual into concessions by the other individual and thus into relative bargaining power.

The role of communication in reaching a consensus in democratic societies has been stressed a great deal in political science (see e.g., Elster (1998)) and philosophy (see e.g., Habermas (1987)). Ultimately, deliberation and consensus require that the arguments and the logical structure of others are accepted as such, and weighted against one's own arguments. The power of voice assumes such an (optimistic) perspective. Each group member views hypothetical outside options as acceptable arguments, and is willing to grant them the weight corresponding to the extent that they can be voiced and articulated logically.

11.3.2 Power of (Un)Friendliness

The power of (un)friendliness also views the household structure, that is, the partition of the population into households, as an endogenous outcome. Moreover, we assume that the level of friendliness is an endogenous individual choice.

The label "friendliness" stands for personal choices like showing a friendly or sour face; choosing a warm or cold tone; paying attention to or ignoring fellow household members. More generally, "friendliness" can serve as a generic term for any personally chosen attribute that exerts a positive externality upon others and increases with the adopted level. Friendliness is neither marketable nor arrangeable by contract. Thus it is neither a standard commodity nor a household-produced commodity, a case already encompassed by the traditional multi-member household model.

Although friendliness is observable, it is in general not contractible, and is left fully to the discretion of each individual. We note that there are important exceptions to that. The degree of friendliness towards customers, for instance, is part of the job description in many service occupations. Moreover, some people may have no choice in the matter. They cannot help being friendly—or obnoxious—either by nature or by habit. Others may be naturally friendly or naturally obnoxious, but are able to act out of character, if they make a conscious effort.

For the sake of simplicity, we assume that the individuals in a household can effortlessly choose to be friendly or unfriendly. Individuals in households, however, often face a trade-off between being nice, understanding and friendly to their partner, and an associated reduction of bargaining power when it comes to the allocation of resources—or surplus in a broad sense—within the household. This conflict arises from the fact that more often than not, enhanced bargaining power derives from a stern and tough posture which sabotages the attempt to appear friendly. We call this phenomenon **the coupling condition**:

Ceteris paribus, a friendlier attitude reduces a person's bargaining power.

The coupling condition rules out the possibility that a person grants the pleasure of friendly company to others and conveys the image of a hard negotiator at the same time. It also rules out the possibility that friendliness is reciprocated, so that friendly behavior triggers a friendly response.[1] Finally, it rules out emotional altruism: The

[1]Basu (1999) explores the idea that while labor supply and consumption of household members are determined by the balance of power within the household (and relative prices), the household's balance of power in turn depends on individually earned income, hence on individual labor supply. In a *household equilibrium*, both the allocation of resources and the balance of power within the household are endogenized simultaneously. Basu (2006) pursues this idea further.

partner's direct benefit from friendly behavior does not contribute to one's own welfare. This is not to say that the assumed away possibilities are unimportant or uninteresting—they can be added in suitable extensions of the concept. Here we accentuate the simplest and most intriguing trade-off.

At a deeper level, there are two justifications why greater friendliness can translate into lower bargaining power. First, studies in social psychology have consistently indicated that negotiators who develop positive attitudes and seek to understand one another's perspective have made greater concessions in negotiations than others. Those studies have been discussed in detail in Gersbach and Haller (2009b).

A second justification can be derived from the contest literature. It is inspired by the work of Che and Gale (1998, 2000). Suppose a two-person household has to decide how to use its resources, and that the decision has to be taken within a limited time-span. Each person has half of the time at its disposal, and has to decide before the bargain starts how it wants to use its allotted time. In particular, a person can use its time to try and understand the perspective and needs of the other person, or to argue its own case. The first type of activity increases the friendliness, while the latter increases the bargaining power.

Prima facie, it appears that the individuals in a society where the coupling condition prevails have no reason to be friendly, since all they would get is a worse bargain within their households. They are penalized for their friendliness. This is certainly correct when a household is considered in isolation. However, competition for partners can make these people friendly. A situation in which a friendly household member is taken advantage of by an unfriendly partner is unstable if the friendly person can find better opportunities outside this household, either by going single or by teaming up with another partner. This suggests that in general, a stable household structure requires friendly behavior by all parties.

Indeed, we show in Gersbach and Haller (2009b) that competition for partners tends to promote friendliness, but can also have a destabilizing effect on households. Moreover, we find that affluence may not buy affection, but that an affluent partner may withhold affection. In general, the degree of friendliness proves sensitive to the socio-economic composition of the population.

11.4 Formal versus Real Power and General Perspective

In a third step, we will examine the relationship between formal power and real power. We will investigate the origin of real power, the issue whether high or maximal real power is compatible with Pareto efficiency, whether higher formal power translates into higher real power, and how such changes in one household affect the utility of members in other households.

For this purpose, we use the notion of a competitive equilibrium with free exit introduced in Chap. 7. We define real power in households in such an equilibrium and identify instances of its absence as well as its presence. The main insights from the pertinent investigation in Gersbach and Haller (2008) are as follows. To begin with, there is a variety of circumstances in which the presence of maximal real power is compatible with Pareto efficiency. That is, high real power *per se* is no indication of social inefficiency. Second, higher formal power does not necessarily translate into higher equilibrium utility or higher real power because groups may dissolve or relative price changes may offset higher formal power. Third, we identify conditions under which changes of formal power in one household do not affect other households in society and, thus, power spillovers are absent. Fourth, we establish existence of competitive equilibria with free exit, including equilibria where some individual enjoys real power. The general question under which circumstances competitive equilibria with free exit exist has been dealt with in Chap. 7.

Overall, our approach enables us to make a first, albeit moderate step toward the study of endogenous power, its determinants and consequences in general equilibrium. This approach is related to the influential work of Hirschman (1970) who has considered the comparative efficiency of the exit and voice options as mechanisms of recuperation. Our analysis suggests that the exit option limits power as long as externalities in groups are sufficiently small.

Finally, our notions of real and formal power can be viewed as a parallel to formal and real authority in organizations (Aghion and Tirole (1997) and Rajan and Zingales (1998)). The former refers to the right to decide and the latter to effective control over decisions. In our model, formal power captures the say in collective decisions and real power captures the increase of an individual's welfare resulting from such decisions. As in the context of formal and real authority, we stress that higher formal power does not necessarily go hand in hand with real power.

11.5 Example: Impact of Power Changes

In essence, this section coincides with Gersbach and Haller (2009a). Proofs are omitted. Section 11.5.5 is condensed.

11.5.1 The Model

We consider our basic model with a fixed household structure as outlined in Chap. 4 with efficient collective household decisions modeled as the outcome of Nash bargaining.

We repeat the essential elements we are using in this section. The population is divided into finitely many households $h = 1, \ldots, n$, with $n \geq 2$. Each household h consists of finitely many members $i = hm$ with $m = 1, \ldots, m(h)$, $m(h) \geq 1$. Put $I = \{hm : h = 1, \ldots, n;\ m = 1, \ldots, m(h)\}$, the finite population of individuals to be considered.

The commodity space is $\mathbb{R}^\ell$ with $\ell \geq 1$. Household h is endowed with a commodity bundle $\omega_h \in \mathbb{R}^\ell$, $\omega_h > 0$. The aggregate or social endowment is $\omega = \sum_h \omega_h$. A generic individual $i = hm \in I$ has:

- consumption set $X_i = \mathbb{R}^\ell_+$;
- preferences $\succsim_i$ on the allocation space $\mathcal{X} \equiv \prod_{j \in I} X_j$ represented by a utility function $U_i : \mathcal{X} \longrightarrow \mathbb{R}$.

The consumption bundle of a generic individual i is denoted by x_i. Let $\mathbf{x} = (x_i)$, $\mathbf{y} = (y_i)$ denote generic elements of $\mathcal{X}$. For $h = 1, \ldots, n$, define $\mathcal{X}_h = \prod_{m=1}^{m(h)} X_{hm}$ with generic elements $\mathbf{x_h} = (x_{h1}, \ldots, x_{hm(h)})$. If $\mathbf{x} \in \mathcal{X}$ is an allocation, then for $h = 1, \ldots, n$, household consumption is given by $\mathbf{x_h} = (x_{h1}, \ldots, x_{hm(h)}) \in \mathcal{X}_h$. There may be intra-household consumption externalities, i.e.:

(E1) Intra-Household Externalities: $U_i(\mathbf{x}) = U_i(\mathbf{x_h})$ for $i = hm$, $\mathbf{x} \in \mathcal{X}$.

A special case is the absence of externalities, i.e.:

(E2) Absence of Externalities: $U_i(\mathbf{x}) = U_i(x_i)$ for $i = hm$, $\mathbf{x} = (x_i) \in \mathcal{X}$.

Budget Constraints: Now consider a household h and a price system $p \in \mathbb{R}^\ell$. For $\mathbf{x_h} = (x_{h1}, \ldots, x_{hm(h)}) \in \mathcal{X}_h$, denote total household expenditure

$$p * \mathbf{x_h} := p \cdot \left(\sum_{m=1}^{m(h)} x_{hm} \right).$$

Then h's **budget set** is defined as $B_h(p) = \{\mathbf{x_h} \in \mathcal{X}_h \;:\; p * \mathbf{x_h} \leq p \cdot \omega_h\}$. We define the **efficient budget set** $EB_h(p)$ by:

$\mathbf{x_h} = (x_{h1}, \ldots, x_{hm(h)}) \in EB_h(p)$ if and only if $\mathbf{x_h} \in B_h(p)$ and there is no $\mathbf{y_h} \in B_h(p)$ such that

$$U_{hm}(\mathbf{y_h}) \geq U_{hm}(\mathbf{x_h}) \quad \text{for all } m = 1, \ldots, m(h);$$
$$U_{hm}(\mathbf{y_h}) > U_{hm}(\mathbf{x_h}) \quad \text{for some } m = 1, \ldots, m(h).$$

General Equilibrium: A **competitive equilibrium (among households)** is a price system p together with an allocation $\mathbf{x} = (x_i)$ satisfying

(i) $\mathbf{x_h} \in EB_h(p)$ for $h = 1, \ldots, n$, and
(ii) $\sum_i x_i = \omega$.

Thus, in a competitive equilibrium among households $(p; \mathbf{x})$, each household makes an efficient choice under its budget constraint and markets clear.[2]

[2]Negative intra-household externalities may imply that a household has a bliss point. If this happens, the social feasibility or market clearing condition (ii) has to be replaced by the **free disposal condition** $\sum_i x_i \leq \omega$.

Nash Bargaining. An efficient household choice under a budget constraint may be the outcome of maximizing a function of the form

$$W_h(\mathbf{x_h}) = S_h(U_{h1}(\mathbf{x_h}), \ldots, U_{hm(h)}(\mathbf{x_h})),$$

subject to the budget constraint. A special case thereof is a **Nash-bargained household decision**. In this case, S_h assumes the form

$$S_h(U_{h1}, \ldots, U_{hm(h)}) = \prod_{m=1}^{m(h)} U_{hm}^{\alpha_{hm}}, \tag{11.1}$$

with the provision that $\alpha_{hm} \geq 0$ and $U_{hm} \geq 0$ for $m = 1, \ldots, m(h)$. The bargaining weight α_{hm} measures the **relative bargaining power** of individual $i = hm$ within household h. In the sequel, we shall concentrate on two-person households, i.e., $m(h) = 2$. We assume $\alpha_{h1}, \alpha_{h2} > 0$ and $\alpha_{h1} + \alpha_{h2} = 1$.

The assumption of Nash-bargained and, hence, efficient household decisions serves us well for the present inquiry into the consequences of shifts of bargaining power. The empirical question of whether collective household decisions are Nash-bargained, indeed, has gotten a fair amount of attention, in particular in the debate between Chiappori (1988, 1991) on the one side and McElroy and Horney (1981, 1990) on the other side (see Bergstrom (1997) for discussions). There has been a growing number of empirical studies performing empirical tests of the collective rationality approach which nests Nash bargaining models as particular cases (Udry (1996); Fortin and Lacroix (1997); Browning and Chiappori (1998); Chiappori et al. (2002), among others).

Two qualifying comments are warranted. First, the interpretation of the maximands of S_h as Nash-bargained outcomes assumes that for each member of a multi-person household, the individual's reservation utility level is zero. The choice of disagreement points for intra-household bargaining is somewhat controversial and depends on the assumed inside or outside options of household members. Therefore, we opt here for a price-independent reservation utility which we normalize to zero solely for computational convenience. However, in Sect. 11.5.5 and in more detail in Sect. 6 of Gersbach and Haller (2009a), we discuss the extension of our model to price-dependent outside options.

Second, although maximization of the Nash product (1) describes the way in which the household reaches an efficient collective decision, it would be a mistake to attribute further meaning to the maximal value of (1) and to changes of it. Normative statements always refer to individuals, either one by one, identifying gainers and losers, or as constituents of society. Pareto-optimality and Pareto-improvements are defined in the standard fashion.

The economies and corresponding examples in this chapter all have locally unique competitive equilibria and possess the budget exhaustion property which stipulates that each household's efficient choices under its budget constraint lie on the household's "budget line". Therefore, as indicated in Chap. 4, equilibrium allocations are Pareto-optimal and comparative statics moves the economy from one Pareto-optimum to another one. Consequently, if a household member gains from a shift in bargaining power, then someone else inside or outside the household must lose.

11.5.2 General Comparative Statics for a Two-Person Household

In this subsection we perform comparative statics with respect to the balance of bargaining power within a two-person household denoted by h. We allow for an arbitrary number of commodities and we consider the general case of intra-household externalities. The entire population consists of an arbitrary number n of households.

Preliminaries. We shall perform comparative statics with respect to the bargaining weights within a select two-person household h, with members $h1$ and $h2$. Whenever convenient and unambiguous, we shall drop the household name and simply refer to consumers 1 and 2. Without restriction, we may also assume that our selected household has the lowest number, i.e., $h = 1$ and the other households are labeled $k = 2, \ldots, n$. For the sake of convenience, we shall further adopt the notation $\alpha = \alpha_{h1}$ and $1 - \alpha = \alpha_{h2}$ so that comparative statics can be performed with respect to the parameter $\alpha \in (0, 1)$. Finally, denote $F \equiv \ln S_h$. We obtain

$$F = F(U_1(\mathbf{x_h}), U_2(\mathbf{x_h}); \alpha) = \alpha \ln U_1(\mathbf{x_h}) + (1 - \alpha) \ln U_2(\mathbf{x_h}). \qquad (11.2)$$

While α is treated as variable, the other characteristics of household h as well as all the characteristics of the rest of the households remain fixed. Each household $k \neq h$ is assumed to choose an efficient consumption plan, $\mathbf{x_k} \in EB(p)$. It may, but need not, maximize a Nash product.

We assume sufficient regularity in the sense that for each $\alpha \in (0, 1)$, the economy has an equilibrium $(p(\alpha); \mathbf{x}(\alpha))$ satisfying:

(iii) local uniqueness and
(iv) continuous differentiability in α.

For each α, at the given price system $p(\alpha)$, household h solves the problem

$$\max\ F(U_1(\mathbf{x_h}), U_2(\mathbf{x_h}); \alpha)\ s.t.\ p(\alpha)[(x_1 + x_2) - \omega_h] \leq 0. \qquad (11.3)$$

The corresponding solution is $\mathbf{x_h}(\alpha) = (x_1(\alpha), x_2(\alpha))$. The budget constraint can be rewritten $\mathbf{x_h} \in B_h(p(\alpha))$. In turn the household budget set $B_h(p(\alpha))$ defines a set $\mathcal{V}(\alpha)$ of **feasible utility allocations** for household h, given the price system $p(\alpha)$:

$$\mathcal{V}(\alpha) \equiv \{(V_1, V_2) \in \mathbb{R}^2 : (V_1, V_2) = (U_1(\mathbf{x_h}), U_2(\mathbf{x_h})) \text{ for some } \mathbf{x_h} \in B_h(p(\alpha))\}.$$

In the sequel, the term **Pareto frontier** refers to the Pareto frontier of $\mathcal{V}(\alpha)$ in the space of utility allocations for the household. In particular, $(U_1(\mathbf{x_h}(\alpha), U_2(\mathbf{x_h}(\alpha))$ lies on the Pareto frontier and solves the problem

$$\max\ F(V_1, V_2; \alpha)\ s.t.\ (V_1, V_2) \in \mathcal{V}(\alpha). \tag{11.4}$$

Finally, for the household under consideration and a given α, the term α-**indifference curve** refers to a locus in $\mathbb{R}^2$ given by an identity $F(V_1, V_2; \alpha) \equiv \text{const.}$

It is instructive to look first at the case $\ell = 1$ of a single good. Assuming that the equilibrium price is positive, the household's budget set and, therefore, its Pareto frontier is price-independent and the household's consumption decision is reduced to the division of a given pie. Consider an increase from α to $\alpha + \epsilon$. Then there are only two possibilities. It can happen that

$$(U_1(\mathbf{x_h}(\alpha)), U_2(\mathbf{x_h}(\alpha))) = (U_1(\mathbf{x_h}(\alpha + \epsilon)), U_2(\mathbf{x_h}(\alpha + \epsilon)))$$

because of a kinked Pareto frontier or a corner solution. But whenever

$$(U_1(\mathbf{x_h}(\alpha)), U_2(\mathbf{x_h}(\alpha))) \neq (U_1(\mathbf{x_h}(\alpha + \epsilon)), U_2(\mathbf{x_h}(\alpha + \epsilon))),$$

consumer 1 benefits from her increased bargaining power to the detriment of consumer 2. This follows from the fact that an increase in 1's bargaining power, that is, in α, renders the household's α-indifference curves steeper.

Bargaining and Price Effects. We focus on binding budget constraints. At the solution $\mathbf{x_h}(\alpha) = (x_1(\alpha), x_2(\alpha))$ of the household's problem (11.4) the equation

$$\frac{\partial F}{\partial U_1} \cdot D_{x_i} U_1 + \frac{\partial F}{\partial U_2} \cdot D_{x_i} U_2 = \lambda(\alpha) p(\alpha), \tag{11.5}$$

with positive Lagrange multiplier $\lambda(\alpha)$ holds for $i = 1, 2$.
For the value function

$$\Phi(\alpha) \equiv F(U_1(\mathbf{x_h}(\alpha)), U_2(\mathbf{x_h}(\alpha)); \alpha), \tag{11.6}$$

we obtain

$$\Phi'(\alpha) = \sum_{i=1}^{2} \left[\frac{\partial F}{\partial U_1} \cdot D_{x_i} U_1 + \frac{\partial F}{\partial U_2} \cdot D_{x_i} U_2 \right] \cdot x_i'(\alpha) + \frac{\partial F}{\partial \alpha}. \tag{11.7}$$

From the budget constraint

$$p(\alpha) \cdot [x_1(\alpha) + x_2(\alpha) - \omega_h] \equiv 0,$$

we obtain

$$p(\alpha)\left[x_1'(\alpha)+x_2'(\alpha)\right]=-p'(\alpha)\left[x_1(\alpha)+x_2(\alpha)-\omega_h\right]. \quad (11.8)$$

Substituting (11.5) and (11.8) into (11.7) yields

$$\Phi'(\alpha)=\frac{\partial F}{\partial \alpha}-\lambda(\alpha)p'(\alpha)[x_1(\alpha)+x_2(\alpha)-\omega_h]. \quad (11.9)$$

Without further qualification, it is impossible to sign $\Phi'(\alpha)$. Under additional assumptions, however, one can gain some detailed insights. To this end, let us decompose the effects of a change of consumer 1's relative bargaining power from α to $\alpha+\epsilon$ into two parts:

1. a **pure bargaining effect** when α is changed to $\alpha+\epsilon$ whereas the price system stays at $p(\alpha)$;
2. a **price effect** when relative bargaining power remains constant at $\alpha+\epsilon$ while the price system adjusts from $p(\alpha)$ to $p(\alpha+\epsilon)$.[3]

In Eq. (11.9), $p'(\alpha)$ reflects the price effect.

Negligible Price Effects. In the remainder of this subsection, we examine the case when the price effect is negligible, i.e., $p'(\alpha)\approx 0$.[4] Then we obtain from Eq. (11.9), up to the negligible price effect:

$$\Phi'(\alpha)=\frac{\partial F}{\partial \alpha}=\ln U_1(\mathbf{x_h}(\alpha))-\ln U_2(\mathbf{x_h}(\alpha)). \quad (11.10)$$

We can exploit the following immediate consequence of Eq. (11.10):

Fact 1 *The value function increases (decreases) in* α, *if* $U_1>U_2$ $(U_1<U_2)$.

However, this result alone does not allow the further conclusion that the utility of at least one household member increases (decreases). A look at a more elementary proof of the fact proves instructive. Namely, let without loss of generality $U_1>U_2>0$ and consider α and ϵ with $0<\alpha<\alpha+\epsilon<1$. Then for sufficiently small ϵ, $\mathbf{x_h}(\alpha)\in B_h(p(\alpha+\epsilon))$ and

$$\begin{aligned}
&[U_1(\mathbf{x_h}(\alpha+\epsilon))]^{\alpha+\epsilon}\cdot[U_2(\mathbf{x_h}(\alpha+\epsilon))]^{1-(\alpha+\epsilon)}\\
&\geq [U_1(\mathbf{x_h}(\alpha))]^{\alpha+\epsilon}\cdot[U_2(\mathbf{x_h}(\alpha))]^{1-(\alpha+\epsilon)}\\
&=[U_1(\mathbf{x_h}(\alpha))]^{\alpha}\cdot[U_2(\mathbf{x_h}(\alpha))]^{1-\alpha}\cdot(U_1/U_2)^{\epsilon}\\
&>[U_1(\mathbf{x_h}(\alpha))]^{\alpha}\cdot[U_2(\mathbf{x_h}(\alpha))]^{1-\alpha}.
\end{aligned}$$

[3]Of course, the price effect could be further decomposed into a substitution and an income effect. But that is immaterial to our analysis.

[4]Note that the price effect vanishes when the budget constraint is not binding.

The last inequality shows that the shift in bargaining power has a "nominal effect" on the household's Nash product even before reoptimization takes place. For this reason, we cannot conclude from a surge of the household's maximum value of F *per se* that the utility of at least one household member has increased. The impact of a shift of bargaining power has to be assessed for each household member individually.

In order to sign individual utility changes, we focus on Eq. (11.5) which is the key to the pure bargaining effect. With $DU_j = (D_{x_1}U_j, D_{x_2}U_j)$ for $j = 1, 2$, let us rewrite (11.5) as

$$\frac{\alpha}{U_1} \cdot DU_1 = -\frac{1-\alpha}{U_2} \cdot DU_2 + \lambda(\alpha)(p(\alpha), p(\alpha)). \tag{11.11}$$

Now consider a change $\Delta\mathbf{x_h}$ away from $\mathbf{x_h}(\alpha)$ while maintaining the budget identity, i.e., $p(\alpha) * (\mathbf{x_h}(\alpha) + \Delta\mathbf{x_h}) = p(\alpha) * \mathbf{x_h}(\alpha) = p(\alpha)\omega_h$. Then $(p(\alpha), p(\alpha)) \cdot \Delta\mathbf{x_h} = p(\alpha) * \Delta\mathbf{x_h} = 0$, hence with (11.11),

$$\left[\frac{\alpha}{U_1} \cdot DU_1\right] \cdot \Delta\mathbf{x_h} = -\left[\frac{1-\alpha}{U_2} \cdot DU_2\right] \cdot \Delta\mathbf{x_h}. \tag{11.12}$$

We first examine local comparative statics. One possibility is $(U_1(\mathbf{x_h}(\alpha)), U_2(\mathbf{x_h}(\alpha))) = (U_1(\mathbf{x_h}(\alpha+\epsilon)), U_2(\mathbf{x_h}(\alpha+\epsilon)))$. The second possibility is $(U_1(\mathbf{x_h}(\alpha)), U_2(\mathbf{x_h}(\alpha))) \neq (U_1(\mathbf{x_h}(\alpha+\epsilon)), U_2(\mathbf{x_h}(\alpha+\epsilon)))$. An increase of α makes the household's α-indifference curves steeper. Hence, as long as $\mathbf{x_h}(\alpha+\epsilon) \in B_h(p(\alpha))$ and $\mathbf{x_h}(\alpha) \in B_h(p(\alpha+\epsilon))$, the revised utility allocation $(U_1(\mathbf{x_h}(\alpha+\epsilon)), U_2(\mathbf{x_h}(\alpha+\epsilon)))$ must lie to the southeast of $(U_1(\mathbf{x_h}(\alpha)), U_2(\mathbf{x_h}(\alpha)))$. Thus consumer 1 benefits from a small increase of her bargaining power to the detriment of consumer 2. The foregoing local comparative statics can be easily globalized.

Proposition 11.1 (Gersbach and Haller (2009a) Proposition 1) *Suppose that the price effect is negligible. If* $0 < \alpha_* < \alpha^* < 1$, *then one of the following two assertions holds:*

(i) $U_1(\mathbf{x_h}(\alpha_*)) = U_1(\mathbf{x_h}(\alpha^*)),\ U_2(\mathbf{x_h}(\alpha_*)) = U_2(\mathbf{x_h}(\alpha^*))$.
(ii) $U_1(\mathbf{x_h}(\alpha_*)) < U_1(\mathbf{x_h}(\alpha^*)),\ U_2(\mathbf{x_h}(\alpha_*)) > U_2(\mathbf{x_h}(\alpha^*))$.

The focus on a particular household h amid many might suggest that shifts of bargaining power are sporadic and therefore price effects are likely to be negligible. Our analysis so far provides valuable insights in case the change of bargaining power is a sporadic event, indeed.

However, drastic price effects will prevail for instance, if the economy is replicated and the same shift in bargaining power occurs in all households that are replicas of h. If the price effect is drastic, both utilities may move in the same direction. The price effect also depends on preferences, including the preferences of consumers not belonging to the household as we will discuss next.

11.5.3 Comparative Statics with Drastic Price Effects

In this section, we allow for drastic price effects and consider a series of examples. The entire population consists of a total of three consumers, two belonging to household h and one forming a one-person household denoted k. To capture widespread shifts in bargaining power in a large finite population, one can consider h as a prototype of a two-person household and k as representative of a one-person household. Literally, one can think in terms of replica economies derived from the basic economies under consideration, with an equal number of two-person households like h and one-person households like k.

Throughout this section, there are always two goods: $\ell = 2$. The second good serves as numéraire. The symbols $x, x_1, x_2, \ldots, x_i, \ldots$ denote quantities of the first good. The symbols $y, y_1, y_2, \ldots, y_i, \ldots$ denote quantities of the second good. c_i^* stands for the equilibrium consumption bundle of a generic person (individual, consumer) i. All consumers fulfill condition E2, i.e., absence of externalities. To simplify the exposition of the later examples, we consider first an auxiliary example of an economy consisting of two one-person households, g and k. The respective consumers are named 0 and 3.

Example 11.0 The initial endowments are $\omega_0 = (1, 0)$ and $\omega_3 = (0, 1)$. The utility representations are

$$u_0 = u_0(x_0, y_0) = x_0^{\alpha} y_0^{1-\alpha}, \text{ with } 0 < \alpha < 1, \text{ and}$$

$$u_3 = u_3(x_3, y_3) = x_3^{1/2} y_3^{1/2}.$$

After normalizing the price of the second good, market equilibrium is unique. The equilibrium price system is

$$p^* = \left(\frac{1}{2(1-\alpha)}, 1\right);$$

the equilibrium consumption bundles are $c_0^* = (\alpha, 1/2)$, $c_3^* = (1-\alpha, 1/2)$. □□

Now we are prepared to consider the case of three individuals, labeled $i = 1, 2, 3$. Consumers 1 and 2 form the two-person household h. In this household, consumer 1 has bargaining power α and consumer 2 has bargaining power $1-\alpha$. Consumer 3 constitutes the single household k. We are going to scrutinize several representative examples which are almost exhaustive in that they exhibit three possible allocative responses to a shift of bargaining power within the two-person household:

(a) Only one member is affected.
(b) The two members are affected in opposite ways.
(c) Both members are affected in the same way.

The examples differ only in individual consumer preferences. The analysis suggests that less substitutability leads to more drastic price effects. We start with the following example of case (a).

Example 11.1 Here consumer 1 benefits from more bargaining power, to the detriment of consumer 3 while consumer 2 is unaffected. Household h is endowed with $\omega_h = (1, 0)$. Its two members, $i = 1, 2$ have utility representations

$$u_1(x_1, y_1) = x_1 \text{ and } u_2(x_2, y_2) = y_2.$$

The household maximizes

$$S_h = u_1^{\alpha} u_2^{1-\alpha} = x_1^{\alpha} y_2^{1-\alpha}, 0 < \alpha < 1.$$

The characteristics of household k are as in the previous example, that is the endowment is $\omega_3 = (0, 1)$ and the utility representation is

$$u_3(x_3, y_3) = x_3^{1/2} y_3^{1/2}.$$

Since the aggregate demand function of household h coincides with the demand function of consumer 0 in Example 11.0, the equilibrium quantities are

$$p^* = \left(\frac{1}{2(1-\alpha)}, 1\right);$$
$$c_1^* = (\alpha, 0), \; c_2^* = \left(0, \frac{1}{2}\right), \; c_3^* = \left(1-\alpha, \frac{1}{2}\right).$$

Hence as asserted consumer 1 benefits from more bargaining power, to the detriment of consumer 3. Consumer 2 is unaffected. □□

In the example, the first good becomes more valuable to the two-person household as the bargaining power of the first consumer increases. This boosts the equilibrium price of the first good and the income of the two-person household endowed with the first good. The household has become richer both in nominal and real terms. Since the expenditure on the second good remains constant, the second consumer is unaffected. But the increase in the residual income to be spent on the first good more than compensates for the higher price: consumer 1 is better off as a consequence of her increased bargaining power. As for consumer 3, his nominal income derived from the possession of the second resource remains constant. Therefore, he has become poorer, has less purchasing power.

From consumer 2's perspective, if bargaining power shifts towards her and prices are fixed, then her welfare is increased. But the resulting price variation offsets her gain. That consumer 2 is unaffected by a change in bargaining power seems to be caused by limited substitutability within the two-person household. This is confirmed by the next example where enhanced bargaining power of consumer 1 translates into improved welfare for this consumer and welfare losses for consumers 2 and 3.

Example 11.2 Here consumer 1 benefits from more bargaining power to the detriment of consumer 2. Consumer 3 either gains or loses. Household h is still endowed with $\omega_h = (1, 0)$. But now each member $i = 1, 2$ has Cobb–Douglas preferences with utility representation

$$u_i(x_i, y_i) = x_i^{\gamma_i} y_i^{1-\gamma_i}, 0 < \gamma_i < 1.$$

The household maximizes

$$u_1^{\alpha} u_2^{1-\alpha} = \left(x_1^{\gamma_1} y_1^{1-\gamma_1}\right)^{\alpha} \left(x_2^{\gamma_2} y_2^{1-\gamma_2}\right)^{1-\alpha}$$
$$= x_1^{\alpha\gamma_1} x_2^{(1-\alpha)\gamma_2} y_1^{\alpha(1-\gamma_1)} y_2^{(1-\alpha)(1-\gamma_2)}.$$

Again, α and $1 - \alpha$ lend themselves as measures of relative bargaining power of consumer 1 and consumer 2, respectively. Household k has the single member 3, with the same consumer characteristics as before. We obtain:

Fact 2 *A shift of bargaining power from consumer 2 to consumer 1 benefits consumer 1 and harms consumer 2, who ends up consuming less of both commodities.*

In Example 11.2 there is more substitutability in the economy than in Example 11.1. □□

Example 11.3 below exhibits less substitutability than Example 11.1, because the preferences of consumer 3 will be altered from Cobb–Douglas to Leontieff. It turns out that the lack of substitution by consumer 3 necessitates a major price adjustment to re-equilibrate the market after bargaining power within household h has shifted. As a result, we observe a very drastic price effect: When bargaining power within their household changes, the equilibrium utilities of consumers 1 and 2 are moving in the same direction.

The example further shows that the aggregate equilibrium consumption of a household can be positively affected by a shift of internal bargaining power. This suggests the possibility that a sophisticated household might succeed in an attempt to manipulate the market outcome, not by misrepresenting endowments or individual preferences, but by misrepresenting the internal bargaining power. To illustrate this novel way of manipulation, which is not yet documented in the literature, suppose the household pretends that the bargaining power of the first consumer is higher than it actually is and they submit the corresponding excess demands to the market. If $\gamma_1 > \gamma_2$, i.e., if the first good is relatively more important to the first consumer, they will end up with a higher aggregate amount of the first good and the same amount of the second good in equilibrium. Whether or not both gain from a successful manipulation depends on the internal distribution of aggregate consumption. If they divided the goods in accordance with their pretended bargaining power, put their money where their mouth is, then consumer 1 would gain and consumer 2 would lose from manipulation. If they divide the goods according to the true bargaining power—

which fixes a proportional sharing rule for each of the goods—then both gain from manipulation.[5]

Example 11.3 Here a shift of bargaining power from consumer 2 to consumer 1 benefits both consumers to the detriment of consumer 3. This example is identical with Example 11.1, except that consumer 3 now has Leontief preferences with utility representation

$$u_3(x_3, y_3) = \min(x_3, y_3).$$

After setting $s = \min(x_3, y_3)$, the utility maximization problem for consumer 3 can be rewritten as

$$\max s \;\; \text{s.t. } (p_1 + 1)s = 1$$

with solution $s = 1/(p_1 + 1)$.

Household h's demand is $(\alpha, (1-\alpha)p_1)$. Therefore, market clearing for the first good requires $1/(p_1 + 1) = 1 - \alpha$. Thus in equilibrium,

$$\begin{aligned} p^* &= (\alpha/(1-\alpha), 1); \\ c_1^* &= (\alpha, 0),\; c_2^* = (0, \alpha),\; c_3^* = (1-\alpha, 1-\alpha). \end{aligned}$$

Thus a shift of bargaining power from consumer 2 to consumer 1 benefits both members of the household to the detriment of consumer 3. A reverse shift harms 1 and 2, and leaves 3 better off. □□

The examples suggest that comparative statics is sensitive to the degree of substitutability in the economy. Enhanced substitutability appears to mitigate price effects. Indeed, if in a further variation of Example 11.1, one assumes linear preferences (perfect substitutability) for consumer 3, with utility representation $u_3(x_3, y_3) = x_3 + y_3$, then the price effect is zero. Moreover, for two-good economies exhibiting CES-utility functions for all individuals with the same elasticity of substitution, the magnitude of the price effect can be parameterized by the elasticity of substitution in the economy. The price effect depends negatively on the elasticity of substitution.

The preceding examples have not contained any sort of externalities. It is straightforward yet space- and time-consuming to analyze the same examples with additive group externalities—where each household member obtains a fixed utility gain from living together—and price-dependent outside options. The qualitative results remain the same.

11.5.4 Comparative Statics Across Households

Until now we have focused primarily on intra-household effects, that is, on the utility changes in a particular household when bargaining power shifts within that house-

---

[5] We will later show that this kind of manipulation is excluded when all individuals have quasi-linear preferences.

hold. Via a series of examples, we have demonstrated that such a shift of bargaining power can affect the members of the corresponding two-person household in three different ways: Only one member is affected; the two members are affected in opposite ways; both members are affected the same way. We have argued earlier that the above examples can be readily reinterpreted as instances of widespread shifts of bargaining power in a replica economy. In the resulting replica economy, the main focus remains on intra-household effects, on the repercussions on the members of those households in which a shift in bargaining power has occurred. However, we have also seen that third parties can be affected. In this section, we redirect our attention to such inter-household or spill-over effects. We start with the quasi-linear case that can serve as a benchmark.

The Quasi-linear Case. We consider a society with $n > 1$ identical households. Household h ($h = 1, \dots, n$) has members $h1$ and $h2$, called the first member and the second member, respectively. There are ℓ goods ($\ell > 1$). The consumption of good k ($k = 1, \dots, \ell$) by individual hi ($i = 1, 2$) is denoted by x^k_{hi}. Each household h is endowed with $w_h = (w^1_h, \dots, w^\ell_h)$. The two members of household h have quasi-linear utility representations of the form

$$U_{h1}(x_{h1}) = u_{h1}\left(x^1_{h1}, \dots, x^{\ell-1}_{h1}\right) + x^\ell_{h1}, \tag{11.13}$$

$$U_{h2}(x_{h2}) = u_{h2}\left(x^1_{h2}, \dots, x^{\ell-1}_{h2}\right) + x^\ell_{h2} \tag{11.14}$$

where u_{hi} is assumed to be strictly concave, strictly increasing and differentiable. Household h maximizes

$$S_h = U_{h1}^{\alpha_h} U_{h2}^{1-\alpha_h} \quad \text{or} \quad ln\, S_h = \alpha_h\, ln\, U_{h1} + (1 - \alpha_h) ln\, U_{h2} \tag{11.15}$$

where $0 < \alpha_h < 1$ is the bargaining power of individual $h1$ in household h. We denote equilibrium values by $\hat{x}^k_{hi}$ and equilibrium utilities by $\hat{U}_{hi}$ and $\hat{u}_{hi}$. For the following we assume that for any array of bargaining power parameters $(\alpha_1, \dots, \alpha_n)$ under consideration, each individual consumes a non-negative amount of the natural numéraire good ℓ in every market equilibrium. We also assume that for any array $(\alpha_1, \dots, \alpha_n)$, the corresponding economy has a unique market equilibrium, up to price normalization. These two assumptions are inessential for our argumentation but simplify the exposition considerably. We shall indicate below which modifications are necessary if the two assumptions are removed. We consider a market equilibrium and parametric changes of the bargaining power in household h and obtain:

Proposition 11.2 (Gersbach and Haller (2009a), Proposition 2) *With quasi-linear preferences:*

(i) A change of α_h in a particular household h does not impact on non-members.

(ii) $\dfrac{\partial \hat{x}^k_{h1}}{\partial \alpha_h} = \dfrac{\partial \hat{x}^k_{h2}}{\partial \alpha_h} = 0$ *for all* $k = 1, \dots, \ell - 1.$

(iii) $\dfrac{\partial \hat{x}^{\ell}_{h1}}{\partial \alpha_h} > 0, \ \dfrac{\partial \hat{x}^{\ell}_{h2}}{\partial \alpha_h} < 0.$

(iv) Suppose that households are homogeneous with respect to individual utility representations and household endowments, with $w_h = \overline{w},\ \forall\, h = 1, \dots, n$. *Then:*

$$\hat{x}^{\ell}_{h1} = \alpha_h \overline{w}^{\ell} + \alpha_h\, \hat{u}_{h2} - (1-\alpha_h)\hat{u}_{h1};$$
$$\hat{x}^{\ell}_{h2} = (1-\alpha_h)\overline{w}^{\ell} + (1-\alpha_h)\, \hat{u}_{h1} - \alpha_h\, \hat{u}_{h2}.$$

The proposition illustrates that with quasi-linear preferences, a change of the bargaining power within a particular household only impacts on the distribution of the numéraire in household h without affecting the consumption of the first $\ell - 1$ commodities. A local change of bargaining power has no price effect and does not affect the utility of individuals in other households. This also means that a household h cannot manipulate outcomes and possibly improve utility of household members at the expense of outsiders by misrepresenting internal bargaining power.

Regarding our simplifying assumptions for the neutrality result, interiority and uniqueness of equilibrium, giving up the first assumption requires to work with Kuhn-Tucker conditions instead of first-order conditions. Without the second assumption, multiple equilibria cannot be ruled out. But a market clearing price system $(p_1, \dots, p_{\ell-1}, 1)$ with respect to some array of bargaining power parameters is also market clearing with respect to all other arrays. Given any such market clearing price system and the associated equilibrium selection, the conclusion of Proposition 11.2 continues to hold.

Individually Preferred Goods. We next turn to situations where internal bargaining power changes in a particular household have spill-over effects on other households. In particular, we examine how individuals are affected if similar (dissimilar) persons in other households can increase their bargaining power. We examine an economy like in the last subsection, but with different individual preferences. We assume households which are homogeneous at the beginning but undergo large sociological changes thereafter. We assume $\ell = 2$ and that all households have the same endowment $w_h = \overline{w} = \left(\overline{w}^1, \overline{w}^2\right) \gg 0$. Individuals in a household h demand different goods, namely:

$$U_{h1}\left(x^1_{h1}, x^2_{h1}\right) = U_{h1}\left(x^1_{h1}\right),$$
$$U_{h2}\left(x^1_{h2}, x^2_{h2}\right) = U_{h2}\left(x^2_{h2}\right).$$

Specifically, individual h1 has a preference for good 1 and demands only that good and h2 has a preference for good 2 and demands only good 2. The utility functions are assumed to be strictly increasing, strictly concave and differentiable. The assumption of household members demanding different goods is one convenient way to divide the society into different sociological groups where individuals are similar within a group and dissimilar across groups. Here we have two groups, "first members" (denoted $h1$) and "second members" (denoted $h2$) of households. Again household h maximizes

$$S_h = U_{h1}^{\alpha_h} U_{h2}^{1-\alpha_h}$$

where $0 < \alpha_h < 1$. We obtain, with $\hat{}$ denoting again equilibrium values:

Proposition 11.3 (Gersbach and Haller (2009a), Proposition 3) *When household members demand different goods, there exists a unique market equilibrium (up to price normalization) for each array* $(\alpha_1, \ldots, \alpha_n)$ *of bargaining power parameters. Moreover, for any two households* $g \neq h$*:*

(i) $\alpha_h > \alpha_g \Rightarrow \hat{x}_{h1}^1 > \hat{x}_{g1}^1$.
(ii) $\alpha_h = \alpha_g \Rightarrow \hat{x}_{h1}^1 = \hat{x}_{g1}^1$.

In the special case where $U_{h1}(x_{h1}^1) = (x_{h1}^1)^{\beta_1}$ and $U_{h2}(x_{h2}^2) = (x_{h2}^2)^{\beta_2}$ with $\beta_1, \beta_2 \in (0, 1)$, further conclusions can be drawn:

(iii) $\partial\hat{x}_{h1}^1/\partial\alpha_h > 0,\ \partial\hat{x}_{g1}^1/\partial\alpha_h < 0$;
(iv) $\partial\hat{x}_{h2}^2/\partial\alpha_h < 0,\ \partial\hat{x}_{g2}^2/\partial\alpha_h > 0$.

Proposition 11.3 has clear-cut implications. Consider the sociological groups "first-members" and "second-members", defined by similarities with respect to preferences. If all individuals in the first sociological group have the same bargaining power (and as a consequence all "second-members" as well), all households consume their endowments since we are in an equilibrium with no active trade. An identical shift of bargaining power across all households has no effect on utilities of any individual either since we will again arrive at an equilibrium with no trade.

The situation is completely different when only some members of a sociological group enjoy higher bargaining power. For instance, a "first-member" suffers when only other "first-members" gain more bargaining power in their respective households. Conversely, the "first-member" benefits from higher own bargaining power as long as other "first-members" do not experience a change of bargaining power. The analogue holds for the other sociological group. Therefore, the main thrust of Proposition 11.3 is that an individual tends to benefit if he can increase his bargaining power but tends to suffer if some or all individuals with the same demand are able to do the same.[6]

For economies of the type discussed in the current subsection, we obtain as an immediate consequence a power illusion phenomenon. Consider two economies denoted by $E_1(\{\alpha_h^1\}_1^n)$ and $E_2(\{\alpha_h^2\}_1^n)$ with households that are homogeneous with respect to individual utility functions and endowments. Household members demand different goods. Equilibrium utilities are denoted by $\hat{U}_{h1}^1, \hat{U}_{h2}^1$ and $\hat{U}_{h1}^2, \hat{U}_{h2}^2$, respectively. Then the following holds:

Corollary 11.1 ***(Power Illusion)***

(i) If $\alpha_h^1 = \overline{\alpha}^1$ *for all* h *and* $\alpha_1^2 > \max_{h\neq1}\{\alpha_h^2\}$, *then* $\hat{U}_{11}^1 < \hat{U}_{11}^2$.

[6]Numerical examples are given in Gersbach and Haller (2009a).

(ii) If $\alpha_h^1 = \overline{\alpha}^1$ *for all* h *and* $\alpha_1^2 < \min_{h \neq 1}\{\alpha_h^2\}$, *then* $\hat{U}_{11}^1 > \hat{U}_{11}^2$.

The corollary illustrates that a member of a sociological group is better off if he has the highest internal bargaining power even if the level of his power is much smaller than in another economy where all individuals of the group have the same bargaining power, that is $\overline{\alpha}^1 > \alpha_1^2$. The underlying intuition runs as follows: Diversity across households opens trade opportunities. The gains from trade will, as a rule, accrue primarily to the members of a sociological group who have relatively higher bargaining power than other members of the group. The absolute level of bargaining power is not important. When, however, the bargaining power of other individuals in the same sociological group is enhanced as well and all individuals of the sociological group end up enjoying an identical level in bargaining power, the original gain is totally eroded.

Separable Utility Function. Of course, individually-preferred goods represent a polar case that yields stark results. Suitable generalizations can weaken these results and introduce further—or opposing—effects. As an example, we consider separable utility functions and examine in which way Proposition 11.3 can be generalized. We again assume $\ell = 2$ and that all households have the same endowment $W_h = \bar{w} = (\bar{w}^1, \bar{w}^2)$. Each household consists of two members: the first member (denoted $h1$) and the second member (denoted $h2$). Individuals in a household h have utility functions

$$U_{h1}(x_{h1}^1, x_{h1}^2) = U_{h1}^1(x_{h1}^1)U_{h1}^2(x_{h1}^2);$$

$$U_{h2}(x_{h2}^1, x_{h2}^2) = U_{h2}^1(x_{h2}^1)U_{h2}^2(x_{h2}^2).$$

The functions $U_{h1}^1(\cdot)$, $U_{h1}^2(\cdot)$, $U_{h2}^1(\cdot)$, and $U_{h2}^2(\cdot)$ are strictly increasing, strictly concave and differentiable. Moreover, the marginal utility of consuming good 1 and good 2 at zero are infinite for both household members. A typical example are Cobb–Douglas utility functions. Household h maximizes

$$S_h = U_{h1}^{\alpha_h} U_{h2}^{1-\alpha_h}$$

where $0 < \alpha_h < 1$.

Such a framework has the following properties. First, suppose all individuals in the first sociological group have the same bargaining power and thus there is no trade. An identical upward shift of their bargaining power will again arrive at an equilibrium with no trade. However, all individuals in the first sociological group benefit from the shift as they will consume more of both goods. Second, spill-overs from an upward shift of bargaining power for a "first-member" in one household can affect other first-members in other households in different ways.

In particular, it is straightforward to construct examples using Cobb–Douglas utility functions with different exponents for first- and second-members where first-members are affected negatively or positively by the change of bargaining power in other households.

11.5.5 Price-dependent Outside Options and Group Externalities

In this subsection, we discuss how the model and results can be extended to price-dependent outside options and group externalities.

The Concept. A simple yet very instructive extension is to introduce the exit option, i.e., the possibility that a household member leaves and forms a single household at the going market prices. Such price-dependent exit options can easily be integrated into our analysis and have been pursued in more detail in Gersbach and Haller (2008) and reported in Gersbach and Haller (2009a). To this end we use the notion **competitive equilibrium with free exit (CEFE)** introduced in Chap. 7. A CEFE is a price system p together with an allocation $\mathbf{x} = (x_i)$ satisfying

(i) $\mathbf{x_h} \in EB_h(p)$ for all $h = 1, \dots, n$.
(ii) $\sum_i x_i = \omega$.
(iii) There are no current household h, household member $i \in h$ and consumption bundle $y_i \in B_{\{i\}}(p)$ such that $U_i(y_i) > U_i(\mathbf{x_h}; h)$.

$U_i(y_i)$ denotes i's utility from consuming the bundle y_i as a single individual. Individual i has an endowment $\omega_{\{i\}}$ and budget set $B_{\{i\}}(p) = \{x_i \in X_i |\; px_i \leq p\omega_{\{i\}}\}$ when forming the single household $\{i\}$. In contrast, $U_i(\mathbf{x_h}; h)$ denotes i's utility when i stays in the multi-member household h and household consumption is $\mathbf{x_h} = (x_j)_{j \in h} \in \mathcal{X}_h$.

In the specific context of **Nash-bargained household decisions**, a CEFE amounts to replacing (11.1) by

$$S_h(U_{h1}, \dots, U_{hm(h)}) = \prod_{m=1}^{m(h)} (U_{hm} - V^o_{hm}(p))^{\alpha_{hm}}$$

where $V^o_{hm}(p)$ is individual hm's indirect utility in the single person household $\{hm\}$ at price system p. This formulation presumes a non-empty set of feasible utility vectors $(U_{h1}, \dots, U_{hm(h)}) \gg (V^o_{h1}(p), \dots, V^o_{hm(h)}(p))$ over which S_h is maximized.

With this formulation, (11.5) remains unchanged while (11.7) becomes

$$\Phi'(\alpha) = \sum_{i=1}^{2} \left[\frac{\partial F}{\partial U_1} \cdot D_{x_i} U_1 + \frac{\partial F}{\partial U_2} \cdot D_{x_i} U_2 \right] \cdot x_i'(\alpha) - \left[\frac{\partial F}{\partial U_1} \cdot \frac{dV^o_1}{dp} + \frac{\partial F}{\partial U_2} \cdot \frac{dV^o_2}{dp} \right] \cdot p'(\alpha) + \frac{\partial F}{\partial \alpha}$$

and (11.9) becomes

$$\Phi'(\alpha) = \frac{\partial F}{\partial \alpha} - p'(\alpha) \left[\frac{\partial F}{\partial U_1} \cdot \frac{dV^o_1}{dp} + \frac{\partial F}{\partial U_2} \cdot \frac{dV^o_2}{dp} \right] - \lambda(\alpha) p'(\alpha) [x_1(\alpha) + x_2(\alpha) - \omega_h].$$

It follows that Proposition 1 generalizes to this case of price-dependent reservation utilities of the form $V_i^o(p)$ where the latter is individuals i's indirect utility in the single person household $\{i\}$ at price system p. A proof of the more general result is omitted. It is analogous to the proof of Proposition 1 but more tedious.

Now let us specialize and consider two-person households $h = \{h1, h2\}$ with **pure group externalities** as follows: Individual hm, $m = 1, 2$ has utility of the form

$$\begin{aligned} U_{hm}(x_{hm}) &= u_{hm}(x_{hm}) && \text{for } x_{hm} \in X_{xm} \text{ when single;} \\ U_{hm}(x_{hm}; h) &= u_{hm}(x_{hm}) + v_m && \text{for } x_{hm} \in X_{xm} \text{ when a member of } h \end{aligned}$$

where $v_m > 0$.

Results. It is straightforward to see that the qualitative properties of Examples 11.1–11.3 remain the same if this type of additive pure group externalities and price-dependent reservation utilities are incorporated. However, the numerical analysis becomes lengthier and more cumbersome.

Next we examine the robustness of the results for quasi-linear preferences and individually preferred goods vis-à-vis this type of additive pure group externalities and price-dependent reservation utilities. Again, the qualitative results tend to remain the same.

Price-dependent outside options limit the feasible allocations in a household as members leave if they are better off as singles. However, as shown in Gersbach and Haller (2008) the first three results of Proposition 11.2 with quasi-linear utilities still hold if outside options are price-dependent. It is obvious that group externalities have the following further consequences. If individual $h1$ gains relatively more from living in household h, i.e., when v_1 increases, he receives less of the numéraire good. But the net effect on utility is positive. Hence, both individuals benefit. The same effects occur when v_2 increases.

Finally, we consider instances of individually preferred goods. With pure group externalities and price-dependent outside options, Proposition 3 and Corollary 1 still obtain, as an inspection of the proof reveals. However, as detailed in Gersbach and Haller (2008), although the qualitative effects of certain changes in bargaining power do not depend on the outside options, the quantitative effects can depend on the specific price-dependent reservation utilities V_{hm}^o.

11.6 Concluding Remarks

The findings reported in the current chapter suggest that our framework naturally lends itself to study power in groups. This first, albeit moderate step warrants further steps towards a deeper understanding of power in groups whose boundaries are movable and which operate in a competitive environment. Besides other conceivable ways to endogenize power in groups, there is also promise to gain insights into the determinants and consequences of formal and real power that individuals or groups might have in society at large. A successful research strategy might combine the ideas of the present chapter with the methodology of Chap. 14.

References

Aghion, P., & Tirole, J. (1997). Formal and real authority in organizations. *Journal of Political Economy, 105*, 1–29.
Basu, K. (1999). Child labor: Causes, consequence, and cure, with remarks on international standards. *Journal of Economic Literature, 37*, 1083–1119.
Basu, K. (2006). Gender and say: A model of household behaviour with endogenously determined balance of power. *The Economic Journal, 116*, 558–580.
Bergstrom, T. C. (1997). A survey of theories of the family. In M. Rosenzweig & O. Stark (Eds.), *Handbook of population and family economics* (pp. 21–79). Amsterdam: Elsevier (Chap. 2).
Browning, M., & Chiappori, P.-A. (1998). Efficient intra-household allocations: A general characterisation and empirical tests. *Econometrica, 66*, 1241–1278.
Che, Y.-K., & Gale, I. (1998). Caps on political lobbying. *American Economic Review, 88*, 643–651.
Che, Y.-K., & Gale, I. (2000). Difference-form contests and the robustness of all-pay auctions. *Games and Economic Behavior, 30*, 22–43.
Chiappori, P.-A. (1988). Nash-bargained household decisions: A comment. *International Economic Review, 29*, 791–796.
Chiappori, P.-A. (1991). Nash-bargained household decisions: A rejoinder. *International Economic Review, 32*, 761–762.
Chiappori, P.-A., Fortin, B., & Lacroix, G. (2002). Marriage market, divorce legislation and household labor supply. *Journal of Political Economy, 110*(1), 37–72.
Elster, J. (Ed.). (1998). *Deliberative democracy*. Cambridge, UK: Cambridge University Press.
Fortin, B., & Lacroix, G. (1997). A test of the unitary and collective models of household labour supply. *Economic Journal, 107*, 933–995.
Gersbach, H., & Haller, H. (2008). Exit and power in general equilibrium. CESifo Working Paper 2369.
Gersbach, H., & Haller, H. (2009a). Bargaining power and equilibrium consumption. *Social Choice and Welfare, 33*, 665–690.
Gersbach, H., & Haller, H. (2009b). The affectionate society: Does competition for partners promote friendliness? *Economic Theory, 40*, 389–403.
Gersbach, H., & Haller, H. (2011). Bargaining cum voice. *Social Choice and Welfare, 36*, 199–225.
Gersbach, H., & Haller, H. (2012). "Hard workers" and labor restrictions. *Journal of Population Economics, 25*, 469–494.
Habermas, J. (1987). *Theory of communicative action*. Boston, MA: Beacon Press.
Hirschman, A. O. (1970). *Exit, voice, and loyalty*. Cambridge, MA: Harvard University Press.
McElroy, M. B., & Horney, M. J. (1981). Nash-bargained household decisions: Toward a generalization of the theory of demand. *International Economic Review, 22*, 333–350.
McElroy, M. B., & Horney, M. J. (1990). Nash-bargained household decisions: Reply. *International Economic Review, 31*, 237–242.
Rajan, R., & Zingales, L. (1998). Power in a theory of the firm. *Quarterly Journal of Economics, 113*, 387–432.
Udry, C. (1996). Gender, agricultural production, and the theory of the household. *Journal of Political Economy, 104*, 1010–1046.

Chapter 12
Risk-sharing Capacity: Markets versus Households

In this chapter we introduce uncertainty in our general equilibrium model with multi-member groups, following the classical state-space approach of Arrow-Debreu. A host of new interesting economic issues emerge. First, risk-averse agents can attempt to insure themselves through markets or through mutual insurance within a multi-member group, say a household, by pooling resources within the group. Which insurance mechanism is chosen and to which extent the mechanisms substitute or complement each other is an open question. Second, one may ask more specifically what is the role of social groups for risk sharing and risk allocation when agents face idiosyncratic or aggregate risk. Third, does a suitable combination of social group formation and contingent commodity markets yield efficient risk allocations? We present a series of examples that shed some light on these issues.

12.1 Introduction

In the sequel, we follow the classical state-space approach of Arrow-Debreu to model uncertainty in general equilibrium theory; see Chap. 7 in Debreu (1957) and Sects. 19 B and 19 C in Mas-Collell et al. (1995). All agents have the same complete description of everything that may happen next period. Contracts and trade in the market are conditional. A contract for the delivery of a commodity specifies, besides its physical characteristics, the state of the world (state of nature) on which the transfer is contingent.

While the formal introduction of uncertainty in our model is straightforward, a host of new interesting economic issues emerge. First, risk-averse agents can attempt to insure themselves through markets or through mutual insurance within a multi-member group, say a household, by pooling resources within the group. Which insurance mechanism is chosen and to which extent the mechanisms substitute or complement each other is an open question. Second, one may ask more specifically

H. Gersbach and H. Haller, *Groups and Markets*,
DOI 10.1007/978-3-319-60516-6_12

what is the role of social groups for risk sharing and risk allocation when agents face idiosyncratic or aggregate risk. Third, does a suitable combination of social group formation and contingent commodity markets yield efficient risk allocations?

The comprehensive examination of these issues is well beyond this chapter and the book. Instead we will present a series of examples that shed some light on these issues. The examples are based on a simple model variant of Gersbach and Haller (2011) extended to situations where endowments are risky. In order to allow for both insurance through markets and insurance through social groups, at least two groups with at least one multi-member group are needed. Insofar, our three-person examples are minimal. They demonstrate that several conceivable scenarios: insurance and risk allocation only through markets, insurance and risk allocation only through households, insurance and risk allocation through both markets and households are possible equilibrium outcomes for particular model parameters.

In all examples, the household structure is exogenously given like in Haller (2000) and Gersbach and Haller (2009). But each example could be modified in such a way that the specific household structure becomes endogenous.

12.2 Set-Up

Let $I = \{1, 2, 3\}$. There are two physical commodities. Moreover, there are two states of the world. We denote by l the goods, by i the individuals, by s the different states, by $w_i^{l,s}$ the endowments, and by $x_i^{l,s}$ the consumption levels. π_s denotes the probability that state s occurs. We simplify the notation by writing $\pi_1 = \pi$ and $\pi_2 = 1 - \pi$.

There are four markets for contingent commodities. The price of commodity l in state s is denoted by p_l^s. The price p_1^1 is normalized to one. Equilibrium prices are denoted by a hat. The endowment bundle for individual i is denoted $\Omega_i = (w_i^{1,1}, w_i^{1,2}, w_i^{2,1}, w_i^{2,2})$.

Preferences for all examples are given as follows. Each individual exhibits pure group externalities and expected utility maximization with respect to consumption. That is, individual i's utility representation assumes the form

$$\mathbb{E}[U_i^{s,c}(x_i^{1,s}, x_i^{2,s})] + U_i^g(h)$$

where the term $U_i^g(h)$ stands for the group externality experienced by the individual if i belongs to group (household) h. More specifically:

$$\mathbb{E}[U_i^{s,c}(x_i^{1,s}, x_i^{2,s})] = \sum_{s=1}^{2} \pi_s \ln(x_i^{i,s}) \qquad \text{for } i = 1, 2;$$

$$\mathbb{E}[U_i^{s,c}(x_i^{1,s}, x_i^{2,s})] = \sum_{s=1}^{2} \pi_s \left(\frac{1}{2}\ln(x_i^{1,s}) + \frac{1}{2}\ln(x_i^{2,s})\right) \qquad \text{for } i = 3;$$

$$U_i^g(h) = \ln(v_i) \text{ with } v_i > 1,$$

$$\text{if } i \in h \text{ and } h = \{1, 2\} \text{ or } h = \{1, 3\};$$

$$U_i^g(h) = 0 \quad \text{otherwise.}$$

We assume that in case $i \in h$ for $h = \{1, 2\}$ or $h = \{1, 3\}$, v_i is sufficiently large so that in all our examples, individuals are better off as members of household $\{1, 2\}$ or household $\{1, 3\}$ than as singletons. This allows us to work with fixed household structures $P = \{\{1, 2\}\{3\}\}$ or $P = \{\{1, 3\}\{2\}\}$ without being concerned whether individuals might be better off by going single.

12.3 General Formulas

We first provide general formulas for a simple set-up that will be used for all numerical examples.

1. Under the household structure $P = \{\{1, 2\}, 3\}$, the optimization problems are given by:

   (i) For the household $\{1, 2\}$,

$$\begin{aligned}&\text{maximize}\\&\Big\{\pi\left(\alpha\ln(x_1^{1,1}) + (1-\alpha)\ln(x_2^{2,1})\right)\\&+ (1-\pi)\left(\alpha\ln(x_1^{1,2}) + (1-\alpha)\ln(x_2^{2,2})\right)\Big\}\\&\text{with respect to } (x_1^{1,1}, x_1^{1,2}, x_2^{2,1}, x_2^{2,2})\end{aligned}$$

   under the budget constraint

$$\begin{aligned}&p_1^1x_1^{1,1} + p_2^1x_2^{2,1} + p_1^2x_1^{1,2} + p_2^2x_2^{2,2}\\=&p_1^1(w_1^{1,1} + w_2^{1,1}) + p_2^1(w_1^{2,1} + w_2^{2,1})\\+&p_1^2(w_1^{1,2} + w_2^{1,2}) + p_2^2(w_1^{2,2} + w_2^{2,2}).\end{aligned}$$

   (ii) For the household $\{3\}$,

$$\max_{\{x_3^{1,1}, x_3^{2,1}, x_3^{1,2}, x_3^{2,2}\}} \Big[\pi\left(\frac{1}{2}\left(\ln(x_3^{1,1}) + \ln(x_3^{2,1})\right)\right) + (1-\pi)\left(\frac{1}{2}\left(\ln(x_3^{1,2}) + \ln(x_3^{2,2})\right)\right)\Big]$$

under the budget constraint

$$p_1^1 x_3^{1,1} + p_2^1 x_3^{2,1} + p_1^2 x_3^{1,2} + p_2^2 x_3^{2,2} = p_1^1 w_3^{1,1} + p_2^1 w_3^{2,1} + p_1^2 w_3^{1,2} + p_2^2 w_3^{2,2}.$$

Which yields the following general solution:

$$\begin{aligned}
x_1^{11}(p) &= \pi\alpha[p_1^2(w_1^{12} + w_2^{12}) + p_2^1(w_1^{21} + w_2^{21}) \\
&\quad + p_2^2(w_1^{22} + w_2^{22}) + w_1^{11} + w_2^{11}], \\
x_1^{12}(p) &= [\alpha(p_1^2(w_1^{12} + w_2^{12}) + p_2^1(w_1^{21} + w_2^{21}) \\
&\quad + p_2^2(w_1^{22} + w_2^{22}) + w_1^{11} + w_2^{11})(1-\pi)]/p_1^2, \\
x_1^{21}(p) &= 0,\ x_1^{22}(p) = 0;
\end{aligned}$$

$$\begin{aligned}
x_2^{11}(p) &= 0,\ x_2^{12}(p) = 0, \\
x_2^{21}(p) &= [(1-\alpha)(p_1^2(w_1^{12} + w_2^{12}) + p_2^1(w_1^{21} + w_2^{21}) \\
&\quad + p_2^2(w_1^{22} + w_2^{22}) + w_1^{11} + w_2^{11})\pi]/p_2^1, \\
x_2^{22}(p) &= [(p_1^2(w_1^{12} + w_2^{12}) + p_2^1(w_1^{21} + w_2^{21}) \\
&\quad + p_2^2(w_1^{22} + w_2^{22}) + w_1^{11} + w_2^{11})(1-\alpha)(1-\pi)]/p_2^2;
\end{aligned}$$

$$\begin{aligned}
x_3^{11}(p) &= \frac{\pi}{2}(1 + p_2^1 w_3^{21} + p_1^2 w_3^{12} + p_2^2 w_3^{22}) \\
x_3^{12}(p) &= \frac{1}{2}\frac{(1-\pi)(1 + p_2^1 w_3^{21} + p_1^2 w_3^{12} + p_2^2 w_3^{22})}{p_1^2}, \\
x_3^{21}(p) &= \frac{1}{2}\frac{(1 + p_2^1 w_3^{21} + p_1^2 w_3^{12} + p_2^2 w_3^{22})\pi}{p_2^1}, \\
x_3^{22}(p) &= \frac{1}{2}\frac{(1 + p_2^1 w_3^{21} + p_1^2 w_3^{12} + p_2^2 w_3^{22})(1-\pi)}{p_2^2}.
\end{aligned}$$

2. Under the household structure $P = \{\{1,3\},2\}$, the optimization problems are given by:

   (i) For the household $\{1,3\}$,

$$\begin{aligned}
&\text{maximize} \\
&\quad \pi\left(\alpha\ln(x_1^{1,1}) + \tfrac{1}{2}(1-\alpha)\left(\ln(x_3^{1,1}) + \ln(x_3^{2,1})\right)\right) \\
&+(1-\pi)\left(\alpha\ln(x_1^{1,2}) + \tfrac{1}{2}(1-\alpha)\left(\ln(x_3^{1,2}) + \ln(x_3^{2,2})\right)\right) \\
&\text{with respect to } (x_1^{1,1}, x_1^{1,2}, x_3^{1,1}, x_3^{2,1}, x_3^{1,2}, x_3^{2,2})
\end{aligned}$$

   subject to the budget constraint

$$
\begin{aligned}
&p_1^1(x_1^{1,1}+x_3^{1,1})+p_2^1x_3^{2,1}+p_1^2(x_1^{1,2}+x_3^{1,2})+p_2^2x_3^{2,2}\\
=&p_1^1(w_1^{1,1}+w_3^{1,1})+p_2^1(w_1^{2,1}+w_3^{2,1})\\
+&p_1^2(w_1^{1,2}+w_3^{1,2})+p_2^2(w_1^{2,2}+w_3^{2,2}).
\end{aligned}
$$

(ii) For the household {2},

$$\max_{\{x_2^{2,1},x_2^{2,2}\}}\left\{\pi\ln(x_2^{2,1})+(1-\pi)\ln(x_2^{2,2})\right\}$$

under budget constraint

$$p_2^1x_2^{2,1}+p_2^2x_2^{2,2}=p_1^1w_2^{1,1}+p_2^1w_2^{2,1}+p_1^2w_2^{1,2}+p_2^2w_2^{2,2}.$$

Which yields the following general solution:

$$
\begin{aligned}
x_1^{11}(p)&=\pi\alpha[p_1^2(w_1^{12}+w_3^{12})+p_2^1(w_1^{21}+w_3^{21})\\
&\quad+p_2^2(w_1^{22}+w_3^{22})+w_1^{11}+w_3^{11}],\\
x_1^{12}(p)&=(1-\pi)\alpha[p_1^2(w_1^{12}+w_3^{12})+p_2^1(w_1^{21}+w_3^{21})\\
&\quad+p_2^2(w_1^{22}+w_3^{22})+w_1^{11}+w_3^{11}]/p_1^2,\\
x_1^{21}(p)&=0,\ x_1^{22}(p)=0;
\end{aligned}
$$

$$
\begin{aligned}
x_2^{11}(p)&=0,\ x_2^{12}(p)=0,\\
x_2^{21}(p)&=\pi\frac{(w_2^{11}+p_1^2w_2^{12}+p_2^1w_2^{21}+p_2^2w_2^{22})}{p_2^1},\\
x_2^{22}(p)&=\frac{(1-\pi)(w_2^{11}+p_1^2w_2^{12}+p_2^1w_2^{21}+p_2^2w_2^{22})}{p_2^2};
\end{aligned}
$$

$$
\begin{aligned}
x_3^{11}(p)&=\frac{1}{2}\pi(1-\alpha)[p_1^2(w_1^{12}+w_3^{12})+p_2^1(w_1^{21}+w_3^{21})\\
&\quad+p_2^2(w_1^{22}+w_3^{22})+w_1^{11}+w_3^{11}],\\
x_3^{12}(p)&=(1-\alpha)(1-\pi)[p_1^2(w_1^{12}+w_3^{12})+p_2^1(w_1^{21}+w_3^{21})\\
&\quad+p_2^2(w_1^{22}+w_3^{22})+w_1^{11}+w_3^{11}]/2p_1^2,\\
x_3^{21}(p)&=\pi(1-\alpha)[p_1^2(w_1^{12}+w_3^{12})+p_2^1(w_1^{21}+w_3^{21})\\
&\quad+p_2^2(w_1^{22}+w_3^{22})+w_1^{11}+w_3^{11}]/2p_2^1,\\
x_3^{22}(p)&=(1-\alpha)(1-\pi)[p_1^2(w_1^{12}+w_3^{12})+p_2^1(w_1^{21}+w_3^{21})\\
&\quad+p_2^2(w_1^{22}+w_3^{22})+w_1^{11}+w_3^{11}]/2p_2^2.
\end{aligned}
$$

12.4 Examples with $P = \{\{1, 2\}\{3\}\}$

In this section, we explore insurance and risk allocation when individuals 1 and 2 form a household.

12.4.1 Example I

In the first example, we consider the following set-up. Endowments $\Omega_1 = (1, 1, 0, 0)$, $\Omega_2 = (0, 1, 1, 0)$, $\Omega_3 = (1, 0, 1, 2)$, household structure $P = \{\{1, 2\}, \{3\}\}$, and $\alpha = 1 - \alpha = \frac{1}{2}$.

This yields the following solution:

- Demand functions are given by:

$$x_1^{11}(p) = \frac{\pi}{2}(1 + p_2^1 + 2p_1^2),\ x_1^{12}(p) = \frac{(1 + p_2^1 + 2p_1^2)(1 - \pi)}{2p_1^2},$$
$$x_1^{21}(p) = 0,\ x_1^{22}(p) = 0;$$

$$x_2^{11}(p) = 0,\ x_2^{12}(p) = 0, x_2^{21}(p) = \frac{(1 + p_2^1 + 2p_1^2)\pi}{2p_2^1},$$
$$x_2^{22}(p) = \frac{(1 + p_2^1 + 2p_1^2)(1 - \pi)}{2p_2^2};$$

$$x_3^{11}(p) = \frac{\pi}{2}(1 + p_2^1 + 2p_2^2),\ x_3^{12}(p) = \frac{(1 + p_2^1 + 2p_2^2)(1 - \pi)}{2p_1^2},$$
$$x_3^{21}(p) = \frac{(1 + p_2^1 + 2p_2^2)\pi}{2p_2^1},\ x_3^{22}(p) = \frac{(1 + p_2^1 + 2p_2^2)(1 - \pi)}{2p_2^2}.$$

- Equilibrium prices are

$$\hat{p}_1^1 = 1,\ \hat{p}_1^2 = \frac{1 - \pi}{\pi},\ \hat{p}_2^1 = 1,\ \hat{p}_2^2 = \frac{1 - \pi}{\pi}.$$

- Equilibrium consumption is

$$\hat{x}_1^{11} = 1,\ \hat{x}_1^{12} = 1,\ \hat{x}_1^{21} = 0,\ \hat{x}_1^{22} = 0;$$
$$\hat{x}_2^{11} = 0,\ \hat{x}_2^{12} = 0,\ \hat{x}_2^{21} = 1,\ \hat{x}_2^{22} = 1;$$
$$\hat{x}_3^{11} = 1,\ \hat{x}_3^{12} = 1,\ \hat{x}_3^{21} = 1,\ \hat{x}_3^{22} = 1.$$

In this example, $\sum_{i=1}^{3} w_i^{l,s} = 2$ for $l = 1, 2$, $s = 1, 2$. Hence, there is no aggregate risk. Moreover, $w_1^{1,1} = w_1^{1,2} = 1$. Hence, as individual 1 does only care about the first commodity (in state 1 and 2), he does not face idiosyncratic risk. Agents 2 and 3 face idiosyncratic risk. Trading in complete markets eliminates idiosyncratic risk altogether and the risk allocation is Pareto efficient. As agent 1 consumes $x_1^{1,1} = x_1^{1,2} = 1$, household formation does not affect trades that occur through markets in equilibrium. Hence, insurance against idiosyncratic risk is solely achieved by contingent commodity markets.

12.4.2 Example II

In the next example we consider the opposite case where insurance occurs solely in the household. Let $\Omega_1 = (0, 1, 0, 1)$, $\Omega_2 = (1, 0, 1, 0)$, $\Omega_3 = (1, 1, 1, 1)$ and as before $P = \{\{1, 2\}, \{3\}\}$. Again $\alpha = 1 - \alpha = \frac{1}{2}$.

We obtain:

- Demand functions are given by:

$$x_1^{11}(p) = \frac{\pi}{2}(1 + p_2^1 + p_1^2 + p_2^2),$$
$$x_1^{12}(p) = \frac{(1 + p_2^1 + p_1^2 + p_2^2)(1 - \pi)}{2p_1^2},$$
$$x_1^{21}(p) = 0,\ x_1^{22}(p) = 0;$$

$$x_2^{11}(p) = 0,\ x_2^{12}(p) = 0,\ x_2^{21}(p) = \frac{(1 + p_2^1 + p_1^2 + p_2^2)\pi}{2p_2^1},$$
$$x_2^{22}(p) = \frac{(1 + p_2^1 + p_1^2 + p_2^2)(1 - \pi)}{2p_2^2};$$

$$x_3^{11}(p) = \frac{\pi}{2}(1 + p_2^1 + p_1^2 + p_2^2),\ x_3^{12}(p) = \frac{(1 + p_2^1 + p_1^2 + p_2^2)(1 - \pi)}{2p_1^2},$$
$$x_3^{21}(p) = \frac{(1 + p_2^1 + p_1^2 + p_2^2)\pi}{2p_2^1},\ x_3^{22}(p) = \frac{(1 + p_2^1 + p_1^2 + p_2^2)(1 - \pi)}{2p_2^2}.$$

- Equilibrium prices are

$$\hat{p}_1^1 = 1,\ \hat{p}_1^2 = \frac{1 - \pi}{\pi},\ \hat{p}_2^1 = 1,\ \hat{p}_2^2 = \frac{1 - \pi}{\pi}.$$

- Equilibrium consumption is

$$\hat{x}_1^{11} = 1,\ \hat{x}_1^{12} = 1,\ \hat{x}_1^{21} = 0,\ \hat{x}_1^{22} = 0;$$
$$\hat{x}_2^{11} = 0,\ \hat{x}_2^{12} = 0,\ \hat{x}_2^{21} = 1,\ \hat{x}_2^{22} = 1;$$
$$\hat{x}_3^{11} = 1,\ \hat{x}_3^{12} = 1,\ \hat{x}_3^{21} = 1,\ \hat{x}_3^{22} = 1.$$

In this example, in equilibrium no trade occurs through markets. Agents in household $\{1, 2\}$ can fully insure themselves by pooling their resources. Markets for contingent commodities prove irrelevant.

12.4.3 Comparison

When we compare Example I and Example II, we observe that $\sum_{i=1}^{3} w_i^{l,s}$ is the same in both examples for both commodities and both states. Furthermore, the allocations coincide. However, in one example (Example I) insurance against idiosyncratic risk is achieved through markets whereas in the second example, this function is performed by the multi-person household. In this respect, markets and households are two equivalent ways to achieve insurance against idiosyncratic risk.

In the next example both markets and households contribute to insurance.

12.4.4 Example III

Endowments $\Omega_1 = (1, 0, 0, 1)$, $\Omega_2 = (0, 0, 1, 1)$, $\Omega_3 = (1, 2, 1, 0)$, household structure $P = \{\{1, 2\}, \{3\}\}$, and $\alpha = 1 - \alpha = \frac{1}{2}$ yield the following solution:

- Demand functions are given by:

$$x_1^{11}(p) = \frac{\pi}{2}(1 + p_2^1 + 2p_2^2),\, x_1^{12}(p) = \frac{(1 + p_2^1 + 2p_2^2)(1 - \pi)}{2p_1^2},$$

$$x_1^{21}(p) = 0,\ x_1^{22}(p) = 0;$$

$$x_2^{11}(p) = 0,\ x_2^{12}(p) = 0,\ x_2^{21}(p) = \frac{(1 + p_2^1 + 2p_2^2)\pi}{2p_2^1},$$

$$x_2^{22}(p) = \frac{(1 + p_2^1 + 2p_2^2)(1 - \pi)}{2p_2^2};$$

$$x_3^{11}(p) = \frac{\pi}{2}(1 + p_2^1 + 2p_1^2),\ x_3^{12}(p) = \frac{(1 + p_2^1 + 2p_1^2)(1 - \pi)}{2p_1^2},$$

$$x_3^{21}(p) = \frac{(1 + p_2^1 + 2p_1^2)\pi}{2p_2^1},\ x_3^{22}(p) = \frac{(1 + p_2^1 + 2p_1^2)(1 - \pi)}{2p_2^2}.$$

- Equilibrium prices are

$$\hat{p}_1^1 = 1,\ \hat{p}_1^2 = \frac{1 - \pi}{\pi},\ \hat{p}_2^1 = 1,\ \hat{p}_2^2 = \frac{1 - \pi}{\pi}.$$

- Equilibrium consumption is

$$\hat{x}_1^{11} = 1,\ \hat{x}_1^{12} = 1,\ \hat{x}_1^{21} = 0,\ \hat{x}_1^{22} = 0;$$
$$\hat{x}_2^{11} = 0,\ \hat{x}_2^{12} = 0,\ \hat{x}_2^{21} = 1,\ \hat{x}_2^{22} = 1;$$
$$\hat{x}_3^{11} = 1,\ \hat{x}_3^{12} = 1,\ \hat{x}_3^{21} = 1,\ \hat{x}_3^{22} = 1.$$

In this example, both markets and households contribute to insurance. The household $\{1, 2\}$ can insure on its own individual 1. However individual 2 cannot be insured by pooling resources in household $\{1, 2\}$. Individual 2 and 3 insure themselves through trading in competitive contingent markets.

In the subsequent examples we consider aggregate risk.

12.4.5 Example IV

Endowments $\Omega_1 = (1, 0, 0, 0)$, $\Omega_2 = (0, 0, 0, 1)$, $\Omega_3 = (1, 2, 1, 0)$, household structure $P = \{\{1, 2\}, \{3\}\}$ and $\alpha = 1 - \alpha = \frac{1}{2}$ yield the following solution:

- Demand functions are given by:

$$x_1^{11}(p) = \frac{\pi}{2}(1 + p_2^2),\ x_1^{12}(p) = \frac{(1 + p_2^2)(1 - \pi)}{2p_1^2},$$
$$x_1^{21}(p) = 0,\ x_1^{22}(p) = 0;$$

$$x_2^{11}(p) = 0,\ x_2^{12}(p) = 0,\ x_2^{21}(p) = \frac{(1 + p_2^2)\pi}{2p_2^1},$$
$$x_2^{22}(p) = \frac{(1 + p_2^2)(1 - \pi)}{2p_2^2};$$

$$x_3^{11}(p) = \frac{\pi}{2}(1 + 2p_1^2 + p_2^1),\ x_3^{12}(p) = \frac{(1 + 2p_1^2 + p_2^1)(1 - \pi)}{2p_1^2},$$

$$x_3^{21}(p) = \frac{(1 + 2p_1^2 + p_2^1)\pi}{2p_2^1},\ x_3^{22}(p) = \frac{(1 + 2p_1^2 + p_2^1)(1 - \pi)}{2p_2^2}.$$

- Equilibrium prices are

$$\hat{p}_1^1 = 1,\ \hat{p}_1^2 = \frac{1 - \pi}{\pi},\ \hat{p}_2^1 = 2,\ \hat{p}_2^2 = 2\frac{1 - \pi}{\pi}.$$

- Equilibrium consumption is

$$\hat{x}_1^{11} = 1 - \frac{\pi}{2},\ \hat{x}_1^{12} = 1 - \frac{\pi}{2},\ \hat{x}_1^{21} = 0,\ \hat{x}_1^{22} = 0;$$
$$\hat{x}_2^{11} = 0,\ \hat{x}_2^{12} = 0,\ \hat{x}_2^{21} = \frac{1}{2} - \frac{\pi}{4},\ \hat{x}_2^{22} = \frac{1}{2} - \frac{\pi}{4};$$
$$\hat{x}_3^{11} = 1 + \frac{\pi}{2},\ \hat{x}_3^{12} = 1 + \frac{\pi}{2},\ \hat{x}_3^{21} = \frac{1}{2} + \frac{\pi}{4},\ \hat{x}_3^{22} = \frac{1}{2} + \frac{\pi}{4}.$$

In this example, we obtain an efficient risk allocation through markets alone. All individuals bear some part of the aggregate risk while all agents consume their desired commodities in all states. Household formation does not contribute to efficient risk sharing: Pooling the resources of individual 1 and 2 does not contribute to risk sharing.

12.4.6 Example V

The next example is given by: Endowments $\Omega_1 = (0, 0, 1, 0)$, $\Omega_2 = (0, 0, 0, 1)$, $\Omega_3 = (1, 1, 1, 1)$ household structure $P = \{\{1, 2\}, \{3\}\}$, and $\alpha = 1 - \alpha = \frac{1}{2}$. It has the following solution:

- Demand functions are given by:

$$x_1^{11}(p) = \frac{\pi}{2}(p_2^1 + p_2^2),\ x_1^{12}(p) = \frac{(p_2^1 + p_2^2)(1 - \pi)}{2p_1^2},$$
$$x_1^{21}(p) = 0,\ x_1^{22}(p) = 0;$$

$$x_2^{11}(p) = 0,\ x_2^{12}(p) = 0,\ x_2^{21}(p) = \frac{(p_2^1 + p_2^2)\pi}{2p_2^1},$$
$$x_2^{22}(p) = \frac{(p_2^1 + p_2^2)(1 - \pi)}{2p_2^2};$$

$$x_3^{11}(p) = \frac{\pi}{2}(1 + p_1^2 + p_2^1 + p_2^2),$$
$$x_3^{12}(p) = \frac{(1 + p_1^2 + p_2^1 + p_2^2)(1-\pi)}{2p_1^2},$$
$$x_3^{21}(p) = \frac{(1 + p_1^2 + p_2^1 + p_2^2)\pi}{4p_2^1},$$
$$x_3^{22}(p) = \frac{(1 + p_1^2 + p_2^1 + p_2^2)(1-\pi)}{2p_2^2}.$$

- Equilibrium prices are

$$\hat{p}_1^1 = 1,\ \hat{p}_1^2 = \frac{1-\pi}{\pi},\ \hat{p}_2^1 = \frac{1}{2},\ \hat{p}_2^2 = \frac{1}{2}\frac{1-\pi}{\pi}.$$

- Equilibrium consumption becomes

$$\hat{x}_1^{11} = \frac{1}{4},\ \hat{x}_1^{12} = \frac{1}{4},\ \hat{x}_1^{21} = 0,\ \hat{x}_1^{22} = 0;$$
$$\hat{x}_2^{11} = 0,\ \hat{x}_2^{12} = 0,\ \hat{x}_2^{21} = \frac{1}{2},\ \hat{x}_2^{22} = \frac{1}{2};$$
$$\hat{x}_3^{11} = \frac{3}{4},\ \hat{x}_3^{12} = \frac{3}{4},\ \hat{x}_3^{21} = \frac{3}{2},\ \hat{x}_3^{22} = \frac{3}{2}.$$

In this example, there is aggregate risk; but the third agent faces no risk. However, the household $\{1, 2\}$ cannot insure itself. Therefore, through trade in contingent commodity markets the third agent partially insures household $\{1, 2\}$ and gets compensated by consuming a large amount of the second commodity.

12.4.7 Example VI

Finally, we consider the following specification: Endowments $\Omega_1 = (1, 0, 1, 1)$, $\Omega_2 = (0, 1, 1, 1)$, $\Omega_3 = (1, 1, 2, 2)$, household structure $P = \{\{1, 2\}, \{3\}\}$, and again $\alpha = 1 - \alpha = \frac{1}{2}$.

This yields the following solution:

- Demand functions are given by:

$$x_1^{11}(p) = \frac{\pi}{2}(1 + 2p_1^2 + p_2^1 + 2p_2^2),$$
$$x_1^{12}(p) = \frac{(1 + 2p_1^2 + p_2^1 + 2p_2^2)(1-\pi)}{2p_1^2},$$
$$x_1^{21}(p) = 0,\ x_1^{22}(p) = 0;$$

$$x_2^{11}(p) = 0,\ x_2^{12}(p) = 0,$$
$$x_2^{21}(p) = \frac{(1 + 2p_1^2 + p_2^1 + 2p_2^2)\pi}{2p_2^1},$$
$$x_2^{22}(p) = \frac{(1 + 2p_1^2 + p_2^1 + 2p_2^2)(1-\pi)}{2p_2^2};$$

$$x_3^{11}(p) = \frac{\pi}{2}(1 + 2p_1^2 + p_2^1 + 2p_2^2),$$
$$x_3^{12}(p) = \frac{(1 + 2p_1^2 + p_2^1 + 2p_2^2)(1-\pi)}{2p_1^2},$$
$$x_3^{21}(p) = \frac{(1 + 2p_1^2 + p_2^1 + 2p_2^2)\pi}{4p_2^1},$$
$$x_3^{22}(p) = \frac{(1 + 2p_1^2 + p_2^1 + 2p_2^2)(1-\pi)}{2p_2^2}.$$

- Equilibrium prices are

$$\hat{p}_1^1 = 1,\ \hat{p}_1^2 = \frac{1-\pi}{\pi},\ \hat{p}_2^1 = \frac{1}{2},\ \hat{p}_2^2 = \frac{1}{2}\frac{1-\pi}{\pi}.$$

- Equilibrium consumption is

$$\hat{x}_1^{11} = 1,\ \hat{x}_1^{12} = 1,\ \hat{x}_1^{21} = 0,\ \hat{x}_1^{22} = 0;$$
$$\hat{x}_2^{11} = 0,\ \hat{x}_2^{12} = 0,\ \hat{x}_2^{21} = 2,\ \hat{x}_2^{22} = 2;$$
$$\hat{x}_3^{11} = 1,\ \hat{x}_3^{12} = 1,\ \hat{x}_3^{21} = 2,\ \hat{x}_3^{22} = 2.$$

In this example, there is aggregate risk. Household $\{1, 2\}$ can achieve some insurance through pooling its resources. However, as the third agent has endowments of the commodities in the same proportion across states like household $\{1, 2\}$, trade in contingent commodity markets cannot improve upon the allocation of risks that is achievable through household formation.

12.5 Examples with $P = \{\{1, 3\}\{2\}\}$

In the examples in this section, we investigate insurance and risk allocation when individuals 1 and 3 form a household.

12.5.1 Example VII

We start again in this section with the following specification: Endowments $\Omega_1 = (1, 1, 0, 0)$, $\Omega_2 = (0, 1, 1, 0)$, $\Omega_3 = (1, 0, 1, 2)$, household structure $P = \{\{1, 3\}, \{2\}\}$, and again $\alpha = 1 - \alpha = \frac{1}{2}$.

This yields the following solution:

- Demand functions are given by:

$$
\begin{aligned}
x_1^{11}(p) &= \frac{1}{2}(2 + p_1^2 + p_2^1 + 2p_2^2)\pi, \\
x_1^{12}(p) &= \frac{(2 + p_1^2 + p_2^1 + 2p_2^2)(1-\pi)}{2p_1^2}, \\
x_1^{21}(p) &= 0, \ x_1^{22}(p) = 0;
\end{aligned}
$$

$$
\begin{aligned}
x_2^{11}(p) &= 0, \ x_2^{12}(p) = 0, \\
x_2^{21}(p) &= \frac{(p_1^2 + p_2^1)\pi}{p_2^1}, \\
x_2^{22}(p) &= \frac{(p_1^2 + p_2^1)(1-\pi)}{p_2^2};
\end{aligned}
$$

$$
\begin{aligned}
x_3^{11}(p) &= \frac{1}{4}(2 + p_1^2 + p_2^1 + 2p_2^2)\pi, \\
x_3^{12}(p) &= \frac{1}{4}\frac{(2 + p_1^2 + p_2^1 + 2p_2^2)(1-\pi)}{p_1^2}, \\
x_3^{21}(p) &= \frac{1}{4}\frac{(2 + p_1^2 + p_2^1 + 2p_2^2)\pi}{p_2^1}, \\
x_3^{22}(p) &= \frac{1}{4}\frac{(2 + p_1^2 + p_2^1 + 2p_2^2)(1-\pi)}{p_2^2}.
\end{aligned}
$$

- Equilibrium prices are

$$
\hat{p}_1^1 = 1, \ \hat{p}_1^2 = \frac{1-\pi}{\pi}, \ \hat{p}_2^1 = \frac{3\pi - 5}{3\pi - 6}, \ \hat{p}_2^2 = \frac{3\pi - 5}{3\pi - 6}\frac{1-\pi}{\pi}.
$$

- Equilibrium consumption is

$$\hat{x}_1^{11} = \frac{4}{3},\ \hat{x}_1^{12} = \frac{4}{3},\ \hat{x}_1^{21} = 0,\ \hat{x}_1^{22} = 0;$$
$$\hat{x}_2^{11} = 0,\ \hat{x}_2^{12} = 0,\ \hat{x}_2^{21} = \frac{6-4\pi}{5-3\pi},\ \hat{x}_2^{22} = \frac{6-4\pi}{5-3\pi};$$
$$\hat{x}_3^{11} = \frac{2}{3},\ \hat{x}_3^{12} = \frac{2}{3},\ \hat{x}_3^{21} = \frac{4-2\pi}{5-3\pi},\ \hat{x}_3^{22} = \frac{4-2\pi}{5-3\pi}.$$

Endowments are as in Example I of the last Section: Agents 2 and 3 face idiosyncratic uncertainty, but $\sum_{i=1}^{3} w_i^{l,s} = 2$ for $l = 1, 2$, $s = 1, 2$, i.e., there is no aggregate risk. In equilibrium, all agents are fully insured. Markets are critical for risk sharing as the second individual has no endowments of the second good in state 2. Household formation, however, is important for the allocation of resources. For instance, individual 1, facing no idiosyncratic risk, achieves higher consumption in both states of the world. This would be impossible if individual 1 were single.

12.5.2 Example VIII

We next consider the following specification: Endowments $\Omega_1 = (0, 0, 1, 1)$, $\Omega_2 = (0, 0, 1, 1)$, $\Omega_3 = (2, 2, 0, 0)$, household structure $P = \{\{1, 3\}, \{2\}\}$, and again $\alpha = 1 - \alpha = \frac{1}{2}$.

- Demand functions are given by:

$$x_1^{11}(p) = \frac{1}{2}(2 + 2p_1^2 + p_2^1 + p_2^2)\pi,$$
$$x_1^{12}(p) = \frac{(2 + 2p_1^2 + p_2^1 + p_2^2)(1-\pi)}{2p_1^2},$$
$$x_1^{21}(p) = 0,\ x_1^{22}(p) = 0;$$

$$x_2^{11}(p) = 0,\ x_2^{12}(p) = 0,$$
$$x_2^{21}(p) = \frac{(p_2^1 + p_2^2)\pi}{p_2^1},$$
$$x_2^{22}(p) = \frac{(p_2^1 + p_2^2)(1-\pi)}{p_2^2};$$

$$x_3^{11}(p) = \frac{1}{4}(2 + 2p_1^2 + p_2^1 + p_2^2)\pi,$$
$$x_3^{12}(p) = \frac{1}{4}\frac{(2 + 2p_1^2 + p_2^1 + p_2^2)(1-\pi)}{p_1^2},$$

$$x_3^{21}(p) = \frac{1}{4}\frac{(2 + 2p_1^2 + p_2^1 + p_2^2)\pi}{p_2^1},$$
$$x_3^{22}(p) = \frac{1}{4}\frac{(2 + 2p_1^2 + p_2^1 + p_2^2)(1-\pi)}{p_2^2}.$$

- Equilibrium prices are

$$\hat{p}_1^1 = 1,\ \hat{p}_1^2 = \frac{1-\pi}{\pi},\ \hat{p}_2^1 = \frac{2}{3},\ \hat{p}_2^2 = \frac{2}{3}.$$

- Equilibrium consumption is

$$\hat{x}_1^{11} = \frac{4}{3},\ \hat{x}_1^{12} = \frac{4}{3},\ \hat{x}_1^{21} = 0,\ \hat{x}_1^{22} = 0;$$
$$\hat{x}_2^{11} = 0,\ \hat{x}_2^{12} = 0,\ \hat{x}_2^{21} = 1,\ \hat{x}_2^{22} = 1;$$
$$\hat{x}_3^{11} = \frac{2}{3},\ \hat{x}_3^{12} = \frac{2}{3},\ \hat{x}_3^{21} = 1,\ \hat{x}_3^{22} = 1.$$

In this example, there is again no aggregate risk. Agents in household $\{1, 3\}$ insure themselves by pooling their resources. In equilibrium no trade occurs through markets. Hence, markets for contingent commodities prove irrelevant. As in the last example, however, individual 1 obtains more consumption goods in the household than by being single.

12.5.3 Example IX

We consider the following specification: Endowments $\Omega_1 = (2, 0, 0, 0)$, $\Omega_2 = (0, 0, 2, 0)$, $\Omega_3 = (0, 2, 0, 2)$, household structure $P = \{\{1, 3\}, \{2\}\}$, and again $\alpha = 1 - \alpha = \frac{1}{2}$.

- Demand functions are given by:

$$x_1^{11}(p) = (1 + p_2^1 + p_2^2)\pi,$$
$$x_1^{12}(p) = \frac{(1 + p_2^1 + p_2^2)(1-\pi)}{p_1^2},$$
$$x_1^{21}(p) = 0,\ x_1^{22}(p) = 0;$$

$$x_2^{11}(p) = 0,\ x_2^{12}(p) = 0,$$
$$x_2^{21}(p) = 2\pi,$$
$$x_2^{22}(p) = \frac{2p_2^1(1-\pi)}{p_2^2};$$

$$x_3^{11}(p) = \frac{1}{2}(1 + p_1^2 + p_2^2)\pi,$$
$$x_3^{12}(p) = \frac{1}{2}\frac{(1 + p_1^2 + p_2^2)(1 - \pi)}{p_1^2},$$
$$x_3^{21}(p) = \frac{1}{2}\frac{(1 + p_1^2 + p_2^2)\pi}{p_2^1},$$
$$x_3^{22}(p) = \frac{1}{2}\frac{(1 + p_1^2 + p_2^2)(1 - \pi)}{p_2^2}.$$

- Equilibrium prices are

$$\hat{p}_1^1 = 1,\ \hat{p}_1^2 = \frac{1-\pi}{\pi},\ \hat{p}_2^1 = \frac{1}{3-3\pi},\ \hat{p}_2^2 = \frac{1}{3\pi}.$$

- Equilibrium consumption is

$$\hat{x}_1^{11} = \frac{4}{3},\ \hat{x}_1^{12} = \frac{4}{3},\ \hat{x}_1^{21} = 0,\ \hat{x}_1^{22} = 0;$$
$$\hat{x}_2^{11} = 0,\ \hat{x}_2^{12} = 0,\ \hat{x}_2^{21} = 2\pi,\ \hat{x}_2^{22} = 2\pi;$$
$$\hat{x}_3^{11} = \frac{2}{3},\ \hat{x}_3^{12} = \frac{2}{3},\ \hat{x}_3^{21} = 2 - 2\pi,\ \hat{x}_3^{22} = 2 - 2\pi.$$

In this example, there is again no aggregate risk and thus all individuals are fully insured. Both markets and households contribute to insurance. The household $\{1, 3\}$ can insure on its own individual 1. However individual 3 cannot be insured by pooling resources in household $\{1, 3\}$. Individual 2 and 3 insure themselves through trading in competitive contingent markets.

In the subsequent examples we consider aggregate risk.

12.5.4 Example X

We consider the following specification: Endowments $\Omega_1 = (0, 0, 0, 1)$, $\Omega_2 = (1, 2, 1, 0)$, $\Omega_3 = (0, 0, 0, 1)$, household structure $P = \{\{1, 3\}, \{2\}\}$ and $\alpha = 1-\alpha = \frac{1}{2}$ yield the following solution:

- Demand functions are given by:

$$x_1^{11}(p) = \pi p_2^2,\ x_1^{12}(p) = \frac{(1-\pi)p_2^2}{p_1^2},$$
$$x_1^{21}(p) = 0,\ x_1^{22}(p) = 0;$$

$$x_2^{11}(p) = 0,\ x_2^{12}(p) = 0,\ x_2^{21}(p) = \frac{(1 + 2p_1^2 + p_2^1)\pi}{p_2^1},$$
$$x_2^{22}(p) = \frac{(1 + 2p_1^2 + p_2^1)(1 - \pi)}{p_2^2};$$

$$x_3^{11}(p) = \frac{p_2^2\pi}{2},\ x_3^{12}(p) = \frac{p_2^2(1 - \pi)}{2p_1^2},$$
$$x_3^{21}(p) = \frac{p_2^2\pi}{2p_2^1}, x_3^{22}(p) = \frac{1 - \pi}{2}.$$

- Equilibrium prices are

$$\hat{p}_1^1 = 1,\ \hat{p}_1^2 = \frac{1 - \pi}{2\pi},\ \hat{p}_2^1 = \frac{4}{3 - 3\pi},\ \hat{p}_2^2 = \frac{2}{3\pi}.$$

- Equilibrium consumption is

$$\hat{x}_1^{11} = \frac{2}{3},\ \hat{x}_1^{12} = \frac{4}{3},\ \hat{x}_1^{21} = 0,\ \hat{x}_1^{22} = 0;$$
$$\hat{x}_2^{11} = 0,\ \hat{x}_2^{12} = 0,\ \hat{x}_2^{21} = \frac{3 + \pi}{4},\ \hat{x}_2^{22} = \frac{3 + \pi}{2};$$
$$\hat{x}_3^{11} = \frac{1}{3},\ \hat{x}_3^{12} = \frac{2}{3},\ \hat{x}_3^{21} = \frac{1 - \pi}{4},\ \hat{x}_3^{22} = \frac{1 - \pi}{2}.$$

In this example, all individuals bear some part of the aggregate risk. Household formation does not contribute to efficient risk sharing since pooling of resources by individual 1 and 3 cannot provide any insurance.

12.5.5 Example XI

We consider the following specification: Endowments $\Omega_1 = (1, 0, 1, 2)$, $\Omega_2 = (0, 0, 1, 2)$, $\Omega_3 = (0, 1, 1, 2)$, household structure $P = \{\{1, 3\}, \{2\}\}$, and again $\alpha = 1 - \alpha = \frac{1}{2}$.

- Demand functions are given by:

$$x_1^{11}(p) = \frac{1}{2}(1 + p_1^2 + 2p_2^1 + 4p_2^2)\pi,$$
$$x_1^{12}(p) = \frac{1}{2}\frac{(1 + p_1^2 + 2p_2^1 + 4p_2^2)(1 - \pi)}{p_1^2},$$
$$x_1^{21}(p) = 0,\ x_1^{22}(p) = 0;$$

$$x_2^{11}(p) = 0,\ x_2^{12}(p) = 0,$$
$$x_2^{21}(p) = \pi + \frac{2p_2^2\pi}{p_2^1},$$
$$x_2^{22}(p) = \frac{(p_2^1 + 2p_2^2)(1-\pi)}{p_2^2};$$

$$x_3^{11}(p) = \frac{1}{4}(1 + p_1^2 + 2p_2^1 + 4p_2^2)\pi,$$
$$x_3^{12}(p) = \frac{1}{4}\frac{(1 + p_1^2 + 2p_2^1 + 4p_2^2)(1-\pi)}{p_1^2},$$
$$x_3^{21}(p) = \frac{1}{4}\frac{(1 + p_1^2 + 2p_2^1 + 4p_2^2)\pi}{p_2^1},$$
$$x_3^{22}(p) = \frac{1}{4}\frac{(1 + p_1^2 + 2p_2^1 + 4p_2^2)(1-\pi)}{p_2^2}.$$

- Equilibrium prices are

$$\hat{p}_1^1 = 1,\ \hat{p}_1^2 = \frac{1-\pi}{\pi},\ \hat{p}_2^1 = \frac{1}{6},\ \hat{p}_2^2 = \frac{1}{12}\frac{1-\pi}{\pi}.$$

- Equilibrium consumption is

$$\hat{x}_1^{11} = \frac{2}{3},\ \hat{x}_1^{12} = \frac{2}{3},\ \hat{x}_1^{21} = 0,\ \hat{x}_1^{22} = 0;$$
$$\hat{x}_2^{11} = 0,\ \hat{x}_2^{12} = 0,\ \hat{x}_2^{21} = 1,\ \hat{x}_2^{22} = 2;$$
$$\hat{x}_3^{11} = \frac{1}{3},\ \hat{x}_3^{12} = \frac{1}{3},\ \hat{x}_3^{21} = 2,\ \hat{x}_3^{22} = 4.$$

In this example, there is again aggregate risk. Household $\{1, 3\}$ can achieve some insurance through pooling its resources. However, since household $\{1, 3\}$ has endowments of the commodities in the same proportion across states as the second agent, trade in contingent commodity markets cannot improve upon the allocation of risks that is achievable through household formation. Therefore, markets for contingent commodities prove irrelevant. We note that the third individual particularly benefits from high consumption of the second commodity since the supply is high in both states of the world, causing low relative prices.

12.6 Concluding Remarks

The preceding examples have highlighted the role of social groups and markets in allocating and diversifying risks in society. To illustrate the scope of possibilities, it suffices to consider the case of two commodities, two states of the world and three individuals. Yet even in that case, a more systematic investigation is left for the future. For instance, one might want to examine further under what circumstances insurance is primarily or exclusively provided by the markets and when it is provided by households. The general case with arbitrary numbers of commodities, states and consumers remains to be studied as well.

Despite our very limited analysis, a few general observations come to mind. First, our examples satisfy budget exhaustion and, therefore, exhibit Pareto optimality of competitive equilibrium allocations. Moreover, in most of the examples, full insurance obtains in equilibrium—which happens because full insurance is both feasible and desirable under the particular numerical specifications. In general, however, full insurance need neither be desirable nor be feasible as illustrated by Example XI.

Second, whether an institution provides insurance may depend on the presence or absence of other insurance opportunities. For instance, the market may not provide insurance because it is impossible or because it is unnecessary. The latter is the case in Example VI. But suppose *ceteris paribus* that household structure $P' = \{\{1\}, \{2\}, \{3\}\}$ prevails. Then insurance through households is no longer possible whereas insurance through markets kicks in. In the opposite direction, insurance through markets may not be available in less developed countries because certain markets do not exist or are inaccessible for part of the population. The lack of insurance through markets and other institutions makes insurance through social groups very attractive and quite likely. Therefore, a sizeable literature deals with the question how well mutual insurance and risk sharing in social groups actually performs. Mazzocco and Saini (2012) develop tests for efficient risk sharing in groups and report on the pertinent literature.

Third, in the context of complete markets for contingent commodities, all prices and trades are determined at the beginning of time and not reconsidered at a later date. If, however, markets reopened at a later date, then the question arises whether consumers would keep their original commitments. This might become an issue for multi-member households if intra-household welfare weights, α and $1 - \alpha$ in our examples, do shift over time. That issue has been addressed empirically by Mazzocco (2007).

Fourth, households play a dual role in our model regarding the allocation of commodities. First, they allow to pool resources and thus may at least partially insure their members. Second, consumption allocation depends on the power of individuals in households, captured in our model by their weights in the utilitarian household welfare function. Individuals may do well in terms of insurance and well (or worse) in terms of actual consumption levels they receive in the household.

Needless to say that provision of insurance is not confined to competitive markets for contingent commodities on the one hand and to mutual insurance by social groups on the other hand. One alternative is publicly provided insurance through social insurance and transfer schemes, for example insurance against illness, unemployment, or severe poverty. Another alternative are contracts, specifying price-quantity combinations, offered by the private insurance industry. These alternative institutions especially are confronted with issues of asymmetric information in the form of adverse selection and moral hazard. Their interaction with the two institutions delineated above, markets and social groups, is intriguing and of utmost interest.

References

Debreu, G. (1957). *Theory of value: An axiomatic analysis of economic equilibrium*. New Haven and London: Yale University Press.

Gersbach, H., & Haller, H. (2009). Bargaining power and equilibrium consumption. *Social Choice and Welfare*, *33*, 665–690.

Gersbach, H., & Haller, H. (2011). Competitive markets, collective decisions and group formation. *Journal of Economic Theory*, *146*, 275–299.

Haller, H. (2000). Household decisions and equilibrium efficiency. *International Economic Review*, *41*, 835–847.

Mas-Collell, A., Whinston, M. D., & Green, J. R. (1995). *Microeconomic theory*. Oxford, UK: Oxford University Press.

Mazzocco, M. (2007). Household intertemporal behavior: A collective characterization and a test of commitment. *Review of Economic Studies*, *74*, 857–895.

Mazzocco, M., & Saini, S. (2012). Testing efficient risk sharing with heterogeneous risk preferences. *American Economic Review*, *102*, 428–468.

Chapter 13
Inefficient Household Decisions

As mentioned in the introduction, a lecture by Pierre-André Chiappori motivated Hans Haller to incorporate multi-member households adhering to collective rationality in a general equilibrium model. Since then, we have assumed collective rationality in most of our joint work. Yet the frequently asked question what would happen if household decisions were inefficient has not totally escaped us. The question actually comprises two queries: Does our analysis depend heavily on the assumption of collective rationality of households? And what can be said, if anything, when the collective rationality assumption is dropped? Assuming collective rationality or efficiency is an obvious starting point when dealing with multi-member households and departing from the unitary model. For it allows to address the question whether the standard results of general equilibrium theory, existence and welfare theorems, still hold when preference maximizing consumers are replaced by collectively rational multi-member households. By and large, the answers turn out to be positive with a fixed household structure. Aggregate excess demand of an economy with collectively rational households has similar properties as the aggregate excess demand of an Arrow-Debreu economy. Hence the prospects of showing existence are very promising. Existence of competitive equilibrium for finite economies with multi-member households has been demonstrated in different ways by Gersbach and Haller (1999) and Sato (2009), and in Chap. 5. *Prima facie*, one might conjecture that validity of the first welfare theorem depends on the particular efficient bargaining protocols employed by households. The basic insight of Haller (2000) is that this is not the case, that Pareto optimal equilibrium allocations result as long as households exhaust their budgets when making efficient consumption decisions for their members. Finally, Gersbach and Haller (2001) obtain a second welfare theorem. The situation proves drastically different if the household structure becomes endogenous. Then both existence and efficiency of competitive equilibria can fail. Hence efficiency at the household level does not necessarily imply efficiency at the economy level.

H. Gersbach and H. Haller, *Groups and Markets*,
DOI 10.1007/978-3-319-60516-6_13

Already Haller (2000) hints at the possibility of strategic behavior in households, but does not pursue it any further. He mentions that—closer in spirit to Chiappori (1992)—a household may adhere to an income-sharing rule of the form

$$\varphi = (\varphi_i)_{i \in h} : \mathbb{R}^\ell \to \mathbb{R}^h$$

that assigns income $\varphi_i(p)$ to household member i when the price system $p \in \mathbb{R}^\ell$ prevails. The domain of φ is possibly only a subset of relevant price systems like the unit simplex in $\mathbb{R}^\ell$. Collectively, the household faces a budget constraint $\sum_{i \in h} \varphi_i(p) = p\omega_h$. Individually, each household member i chooses

$$x_i^* \in \arg\max\{U_i(x_i, x_{h\setminus i}^*) | \, x_i \in X_i, \, px_i \leq \varphi_i(p)\}$$

given the other members' choices, $x_{h\setminus i}^*$. Thus, given prices and an income-sharing rule, the household members play a Nash equilibrium of a strategic game among themselves. Existence of Nash equilibrium is shown in Browning et al. (2010). Lechene and Preston (2011) perform demand analysis for a "non-cooperative" two-person household $h = \{A, B\}$ where the set of commodities $\{1, \ldots, \ell\}$ is divided into private goods not causing any externalities and local public goods and disposable incomes y^A and y^B are exogenously given.

In the absence of externalities and with individual nonsatiation, the household choice resulting from Nash equilibrium play is efficient. Haller (2000) further shows that under the same conditions, each equilibrium allocation $\mathbf{x}^*$ under collective rationality can be individually decentralized with suitable income shares. Namely, let p^* denote the equilibrium price system. Then $x_h^* \in EB_h(p^*)$ and $p^* * \mathbf{x_h}^* = p^*\omega_h$ for each household h. Let $i \in h$ and $x_i \in X_i$ with $p^*x_i \leq p^*x_i^*$. Then $(x_i, x_{h\setminus i}^*) \in B_h(p^*)$. If $U_i(x_i) > U_i(x_i^*)$, then $\mathbf{x}^* \notin EB_h(p^*)$, a contradiction. Hence $x_i^* \in \arg\max\{U_i(x_i) | \, x_i \in X_i, p^*x_i \leq p^*x_i^*\}$ and income shares $\varphi_i(p^*) = p^*x_i^*$ will do.

It is well known that Nash equilibria are not necessarily efficient. And indeed, household choice based on an income-sharing rule need not be efficient in the presence of externalities. Still, the combination of externalities, income-sharing rules, and selfish individual behavior does not always impede efficiency. For instance, in the presence of certain externalities, selfish individual consumption choices constrained by an income-sharing rule and followed by a welfare-maximizing redistribution of utilities by a benevolent head of household can lead to an efficient household choice as the subgame-perfect equilibrium outcome. For details, see Bergstrom's (1989) investigation of Becker's (1978, 1981) "rotten kid theorem". Nevertheless, household decisions based on an income-sharing rule tend to be inefficient in the presence of externalities.

In Gersbach and Haller (2005, 2006), we study the general equilibrium implications of inefficient household decisions and address the question what can be said if the collective rationality assumption is dropped. We find that inefficiencies at the household level may but need not lead to inefficiency at the economy level: Inefficiency may beget efficiency. We resort to an analytic decomposition of collective household decisions into two steps, which proves convenient for our purposes—regardless whether the household views its decision making that way. First, at the prevailing price system p, household h chooses a net trade $z_h \in \mathbb{R}^\ell$ with the market such that $\overline{x}_h = \omega_h + z_h \geq 0$ and $pz_h \leq 0$. In a more graphic description, it fixes the dimensions of an Edgeworth Box for the household. In the second step, the available aggregate household consumption $\overline{x}_h$ is distributed among the household members, that is, the household chooses $\mathbf{x_h} = (x_i)_{i\in h} \in \mathcal{X}_h$ such that $\sum_{i\in h} x_i = \overline{x}_h$. More graphically, the household picks a point (an allocation) within its previously chosen Edgeworth Box.

Based on this decomposition, two specific types of inefficient household decisions can be isolated. Namely, in order to arrive at an efficient consumption decision under its budget constraint, the household must first choose the right Edgeworth Box and then pick a point on the contract curve in that Edgeworth Box. Therefore, one can identify two sources of inefficiencies committed by the household:

Inefficient net trade with the market: The first type of household inefficiency results from an inefficient net trade with the market and does not rule out global efficiency.

Inefficient internal distribution: The second type of household inefficiency results from an inefficient distribution of the household's aggregate consumption to individual household members and always causes global inefficiency.

Of course, the two types of inefficient decision making can be compounded. But it is analytically convenient to consider each of them separately. More importantly, this sort of piecemeal analysis renders interesting results already.

To formalize the two types of household inefficiency, it is convenient to introduce yet another distinguished subset of a household's budget set. For each household h and every price system p, we define the **potentially efficient budget set** $PEB_h(p)$ as the set of $\mathbf{x_h} = (x_i)_{i\in h} \in B_h(p)$ for which there exists $\mathbf{x'_h} = (x'_i)_{i\in h} \in EB_h(p)$ such that $\sum_{i\in h} x'_i = \sum_{i\in h} x_i$ and $U_i(\mathbf{x'_h}) \geq U_i(\mathbf{x_h})$ for all $i \in h$. When choosing an element from its potentially efficient budget set, the household makes an efficient net trade, but may not achieve efficient internal distribution.

13.1 Inefficient Net Trades

Suppose that a household performs an inefficient net trade with the market which means that the household could improve (in a weak sense) the welfare of its members by making a different choice under its budget constraint, but in order to achieve that would have to change its net trade with the market. That is, the household chooses $\mathbf{x_h} \in B_h(p)\backslash PEB_h(p)$. If the household wants to correct its mistake after market clearing, then the net trades of some other households would have to be altered as well, possibly to the detriment of the welfare of the other households' members. This line of argument suggests that inefficient net trades might lead to an efficient market allocation. To demonstrate this possibility, it suffices to consider the simplest case of two commodities, $\ell = 2$, and a single consumer i. Let the consumer be endowed with the commodity bundle $\omega_i = (1, 1)$ and his preferences be represented by the Cobb-Douglas utility function

$$U_i(x_i) = x_i^1 x_i^2$$

for $x_i = (x_i^1, x_i^2) \in \mathbb{R}_+^\ell$. Suppose that at the price system $p = (1, 2)$, the consumer chooses the net trade $z_i = 0$. Then the market is cleared and $x_i = \omega_i$ is a Pareto-optimal allocation for the one-person economy. But under his budget constraint, the consumer performs an inefficient net trade with the market, because his actual demand $\omega_i = (1, 1)$ differs from his Marshallian demand $x_i(p) = (3/2, 3/4)$. However, the former is Pareto-optimal whereas the latter is socially infeasible. This proves the point. The example can be extended to more commodities, multiple consumers, and varying household sizes. The phenomenon of inefficient net trades leading to Pareto optimal allocations has been more systematically investigated in Gersbach and Haller (2006).

13.2 Inefficient Internal Distribution

Suppose that a household performs an efficient net trade with the market which means that the household can achieve an efficient choice under its budget constraint by suitably dividing its aggregate consumption bundle among its members. But the actually chosen internal distribution of commodities may be inefficient in the sense that redistribution within the household can improve the welfare of its members. If so, the mistake can be rectified simply by internal reallocation without affecting the welfare of members of other households. This leads to the conclusion that inefficient internal distribution, a particular type of inefficient household decision, always begets global inefficiency. Indeed, let the household structure P be given and $(p, \mathbf{x}; P)$ be a state of the economy such that $(\mathbf{x}; P)$ is a feasible allocation. If $h \in P$ is a household with $\mathbf{x_h} \in PEB_h(p) \setminus EB_h(p)$, then there exists $\mathbf{z_h} \in EB_h(p)$ with $\sum_{i \in h} z_i = \sum_{i \in h} x_i$ and $U_i(\mathbf{z_h}; h) \geq U_i(\mathbf{x_h}; h)$ for all $i \in h$. Since $\mathbf{z_h} \in EB_h(p)$, but $\mathbf{x_h} \notin EB_h(p)$, $U_i(\mathbf{z_h}; h) > U_i(\mathbf{x_h}; h)$ has to hold for some $i \in h$. Now set $\mathbf{x'_h} = \mathbf{z_h}$ and $\mathbf{x'_k} = \mathbf{x_k}$ for households $k \neq h$. This defines a feasible allocation $(\mathbf{x}' = (x'_i)_{i \in I}; P)$. Now $U_i(\mathbf{x'_h}; h) > U_i(\mathbf{x_h}; h)$ for certain members i of household h and $U_j(\mathbf{x'_{P(j)}}; P(j)) = U_j(\mathbf{x_{P(j)}}; P(j))$ for all other consumers j. Hence $\mathbf{x}$ is not Pareto-optimal given P.

13.3 When Outside Options Beget Efficiency

Endogenous household formation, availability of outside options and the associated competition for partners can limit—or in some cases prevent—inefficiencies at the household level. Notice that inefficient internal distribution on the part of households constitutes the analogue of technological inefficiency in the production sector. It is a time honored theme in industrial economics that increased competition among producers reduces both allocative and technical inefficiencies.[1] Moreover, potential competition may suffice to further efficiency. To quote Schumpeter (1975, p. 85):

> It is hardly necessary to point out that competition of the kind we now have in mind acts not only when in being but also when it is merely an ever-present threat. It disciplines before it attacks.

In a similar vein, the concept of contestable markets forwarded by Baumol et al. (1982) postulates that potential hit-and-run competition has the same effect as actual competition.

We apply the idea that competitive forces can serve as a disciplinary device to the consumption sector. The hope is that competition will cause the elimination or reduction of inefficient internal distribution in households in a similar manner as it causes erosion of managerial slack in firms. Yet we know from the previous section that competition for resources alone will be to no avail in this respect. However, it turns out that if household stability is threatened by inefficient internal distribution, if in a sense households are competing for resources and members, then the households which exist in equilibrium must make efficient or not too inefficient decisions. This presumes that dissatisfied household members have the option to leave and that household stability (requiring that nobody wants to exercise the option) is an additional equilibrium condition. Accordingly, we are going to investigate whether and to what extent inefficient household decisions due to inefficient internal distribution are sustainable in equilibrium, if individuals have the option to form new and potentially more efficient households.

We find that the exit and joining option alone may not suffice to eliminate all inefficiencies. In Gersbach and Haller (2005), we present an example with a state $(p, \mathbf{x}(\epsilon); P)$ such that

(a) $(\mathbf{x}(\epsilon); P)$ is a feasible allocation;
(b) $\mathbf{x_h}(\epsilon) \in B_h(p) \backslash EB_h(p)$ for all $h \in P$;
(c) in state $(\mathbf{x}(\epsilon), p; P)$, no consumer can benefit from exit;
(d) in state $(\mathbf{x}(\epsilon), p; P)$, no consumer can benefit from joining another household;
(e) the allocation $(\mathbf{x}(\epsilon); P)$ fails to be Pareto optimal because of inefficient internal distribution in each household

where the parameter $\epsilon \in (0, 1)$ measures the degree of inefficiency. Although inefficiencies within households cannot be completely ruled out in the example, the exit option limits the degree of inefficiency a household can afford: In order to prevent a consumer from leaving, the degree of inefficiency cannot be too large. More specifically, a consumer would benefit from exit in case $\epsilon^2 > b$ where $b > 0$ stands for the pure group externality from pairwise matching that prevails in the example. Therefore, $\epsilon^2 \leq b$ has to hold to prevent exit.

---

[1] Leibenstein's much acclaimed 1966 article has raised the awareness for technological inefficiencies or X-inefficiencies. Hart (1983) formalizes the idea that competition in the product market reduces managerial slack.

We finally find that if competition for partners is fierce enough, then inefficient household decisions cannot occur:

Proposition 13.1 (Gersbach and Haller (2005), Proposition 3) *Suppose IPR and GSE and that for each $i \in I$, preferences are continuous, convex and strictly monotone in own consumption. Let $(p, \mathbf{x}; P)$ be a state such that*

(i) $(\mathbf{x}; P)$ *is a feasible allocation;*
(ii) $\mathbf{x_h} \in B_h(p)$ *for all* $h \in P$*;*
(iii) no group benefits from forming a new household.

If g and h are two multi-member households in P of equal size, that is $|g| = |h| > 1$*, then* $\mathbf{x_g} \in EB_g(p)$ *and* $\mathbf{x_h} \in EB_h(p)$.

The proof relies on condition (iii) which means for instance that a member of a two-person household and member of another two-person household should not benefit from forming a new two-person household, and likewise for larger households of equal size. If like in the foregoing example, a household member can only leave and go single or join another household, then the argument does not go through. Both the last proposition and the example could be reformulated in terms of type economies, at the cost of additional notation. Consumer preferences then depend on household profile (number of each type present) rather than household size. In the proposition, the condition of equal household size has to be replaced by equal household profile. In the alternative example, one obtains a simple model of bilateral matching or a "marriage model", if there are two types (male and female) and consumers prefer heterogeneous two-person households to other households.

13.4 The Impact of Production

Here we present an example that shows how production interacts with inefficient household decisions. In the example, some households are prone to make inefficient net trades. In the case of pure exchange, those households make efficient net trades at equilibrium prices—although they would make inefficient net trades at non-equilibrium prices—and the equilibrium allocation is efficient. After a linear production sector is introduced, a different equilibrium price system emerges, the households prone to make inefficient net trades do so at the new prices, and the equilibrium allocation is inefficient.

13.4.1 Consumer and Household Characteristics

We consider an economy with a population of H identical households, denoted $h = 1, \ldots, H$. Each household consists of two members so there is a total of $N = 2H$

consumers and a fixed household structure P. We concentrate on a typical household h with two individuals $i = 1, 2$.

There are two commodities, denoted by $k = 1, 2$. Preferences of an agent i are represented by

$$U_1(\mathbf{x_h}) = \ln x_1^1 + \ln x_1^2 + \beta \ln x_2^2 + v_1$$
$$U_2(\mathbf{x_h}) = \ln x_2^1 + \ln x_2^2 + v_2,$$

where x_i^k denotes the quantity of good k ($k = 1, 2$) consumed by individual i. Individual 1 experiences positive externalities from consumption of individual 2. The strength of these externalities is expressed by the parameter β ($\beta > 0$). Moreover, the variables v_1 and v_2 ($v_1 \geq 0, v_2 \geq 0$) stand for the extent of group externalities that individuals experience when they live together. Finally, individual endowments are given by $\omega_1 = (1, 1), \omega_2 = (1, 1)$ and the household endowment as $\omega_h = (2, 2)$.

We explore three scenarios of this economy: efficient household decisions and pure exchange, inefficient household decisions and pure exchange and inefficient household decisions with production possibilities and exchange. For all three scenarios we normalize the price of the first good to 1.

13.4.2 Efficient Household Decisions and Pure Exchange

Efficient household decisions can be described by the maximization of a utilitarian household welfare function:

$$\max_{\{x_i^k\}_{i=1,2}^{k=1,2}} \{\alpha U_1(\mathbf{x_h}) + (1-\alpha) U_2(\mathbf{x_h})\}$$
$$\text{s.t.} \quad p_2(x_1^2 + x_2^2) + x_1^1 + x_2^1 = 2p_2 + 2.$$

The parameter α ($0 < \alpha < 1$) is the weight of individual 1 in the welfare of household h. We note that group externalities do not affect excess demand of the household. We obtain the following first-order conditions.

$$\frac{\alpha}{x_1^1} - \lambda = 0,$$
$$\frac{\alpha}{x_1^2} - \lambda p_2 = 0,$$
$$\frac{1-\alpha}{x_2^1} - \lambda = 0,$$
$$\frac{\alpha\beta}{x_2^2} + \frac{1-\alpha}{x_2^2} - \lambda p_2 = 0,$$

where λ is the Lagrange multiplier associated with the budget constraint.

This yields:

$$\begin{aligned} x_1^1 &= p_2 x_1^2, \\ (1-\alpha)x_1^1 &= \alpha x_2^1, \\ (1-\alpha+\alpha\beta)x_1^1 &= p_2 \alpha x_2^2 \end{aligned}$$

and

$$\begin{aligned} x_1^2 &= \frac{1}{p_2} x_1^1, \\ x_2^1 &= \frac{1-\alpha}{\alpha} x_1^1, \\ x_2^2 &= \frac{1-\alpha+\alpha\beta}{\alpha p_2} x_1^1. \end{aligned}$$

With the budget constraint we obtain

$$x_1^1 \left(1 + \frac{1-\alpha}{\alpha} + 1 + \frac{1-\alpha+\alpha\beta}{\alpha}\right) = 2p_2 + 2,$$

which yields

$$\begin{aligned} x_1^1 &= \frac{2\alpha(p_2+1)}{2+\alpha\beta}, \\ x_1^2 &= \frac{2\alpha(p_2+1)}{p_2(2+\alpha\beta)}, \\ x_2^1 &= \frac{2(1-\alpha)(p_2+1)}{2+\alpha\beta}, \\ x_2^2 &= \frac{2(1-\alpha+\alpha\beta)(p_2+1)}{p_2(2+\alpha\beta)}. \end{aligned}$$

The excess demand vector of household h, denoted by z_h is given by

$$z_h = \left(\frac{2(p_2+1)}{2+\alpha\beta} - 2, \frac{2(1+\alpha\beta)(p_2+1)}{p_2(2+\alpha\beta)} - 2\right). \tag{13.1}$$

As all households are identical, the market clearing price is

$$p_2^* = 1 + \alpha\beta,$$

which yields the P-optimal commodity allocation

$$\begin{aligned} x_1^{1*} &= 2\alpha \\ x_1^{2*} &= \frac{2\alpha}{1+\alpha\beta} \\ x_2^{1*} &= 2(1-\alpha) \\ x_2^{2*} &= \frac{2(1-\alpha+\alpha\beta)}{1+\alpha\beta}. \end{aligned}$$

13.4.3 *Inefficient Household Decisions*

We next assume that households $h = 1, \dots, H'$ (with $1 \leq H' \leq H$) are prone to make inefficient decisions regarding their net trades. Suppose in particular that those households choose excess demand vectors

$$\tilde{z}_h = \left(t\left(\frac{2(p_2+1)}{2+\alpha\beta} - 2\right), t\left(\frac{2(1+\alpha\beta)(p_2+1)}{p_2(2+\alpha\beta)} - 2\right)\right)$$

with $0 \leq t < 1$. We note that these choices exhaust the budget of the household.

We observe that the market clearing price in the economy remains the same at $p_2^* = 1 + \alpha\beta$. Moreover, the resources each household has at its disposal remain $(2, 2)$. As long as the households internally allocate these resources according to

$$\begin{aligned} x_1^1 &= 2\alpha \\ x_1^2 &= \frac{2\alpha}{1+\alpha\beta} \\ x_2^1 &= 2(1-\alpha) \\ x_2^2 &= \frac{2(1-\alpha+\alpha\beta)}{1+\alpha\beta} \end{aligned}$$

the ensuing consumption allocation is P-optimal. Notice that all households choose the efficient net trade $z_h = (0, 0)$ at the equilibrium price system $(1, p_2^*)$ whereas a subset of households would choose inefficient trades with the market at price systems $(1, p_2)$ with $p_2 \neq p_2^*$.

13.4.4 *Inefficient Household Decisions and Production*

We next show that *ceteris paribus* inefficiencies cannot be avoided when there is a particular kind of production in the economy. This is most easily demonstrated for a linear production function. But the argument extends to other convex production sets.

Suppose that the production function is

$$y_2 = \gamma y_1$$

with $\gamma > 1$. Production is operated by a firm of which each household owns a share $1/H$. The firm uses the amount y_1 of the first commodity to produce quantity y_2 of the second commodity. We note that the equilibrium profit of the firm is zero.

We consider again a scenario such that households $h = 1, \ldots, H'$ (with $1 \leq H' \leq H$) choose inefficient net trades as before while the remaining households choose efficient net trades. Assume that households receive zero dividends—which is the case in equilibrium. We can distinguish two cases where $p_2^* = 1 + \alpha\beta$ is the equilibrium price from above.

CASE 1: $p_2^* \leq 1/\gamma$. Then the equilibrium price system is again $(1, p_2^*)$, the firm chooses $y_1 = y_2 = 0$, equilibrium consumption is as in pure exchange, and the allocation is optimal given the household structure P.

CASE 2: $p_2^* > 1/\gamma$. Then the equilibrium price system is $(1, 1/\gamma)$. For those households that optimize correctly, their equilibrium excess demand vector is given by (13.1) and thus

$$\begin{aligned}
\hat{x}_1^1 &= \frac{2\alpha(1/\gamma + 1)}{2 + \alpha\beta}, \\
\hat{x}_1^2 &= \frac{2\alpha(1/\gamma + 1)}{(1/\gamma)(2 + \alpha\beta)} \\
&= \frac{2\alpha(\gamma + 1)}{2 + \alpha\beta}, \\
\hat{x}_2^1 &= \frac{2(1 - \alpha)(1/\gamma + 1)}{2 + \alpha\beta}, \\
\hat{x}_2^2 &= \frac{2(1 - \alpha + \alpha\beta)(1/\gamma + 1)}{(1/\gamma)(2 + \alpha\beta)} \\
&= \frac{2(1 - \alpha + \alpha\beta)(\gamma + 1)}{2 + \alpha\beta}.
\end{aligned}$$

For households that choose inefficient net trades, the equilibrium excess consumption vector is

$$\hat{z}_h = \left(t \left(\frac{2(1/\gamma + 1)}{2 + \alpha\beta} - 2 \right), t \left(\frac{2(1 + \alpha\beta)(1 + \gamma)}{2 + \alpha\beta} - 2 \right) \right)$$

with $0 \leq t < 1$.

It is now readily verified that if *ceteris paribus* t is sufficiently small and γ is sufficiently large, then the H' households with inefficient net trades are worse off compared to the other households even if they distribute the received resources efficiently within the household. This is obvious for $t = 0$: In the equilibrium consumption plan of efficient households, $\hat{\mathbf{x}}_h = (\hat{x}_1^1, \hat{x}_1^2; \hat{x}_2^1, \hat{x}_2^2)$, for both consumers, consumption of the second good goes to infinity as γ tends to infinity whereas consumption of the first good is bounded away from zero. The equilibrium consumption of inefficient households is $\mathbf{x}_\mathbf{h}^* = (x_1^{1*}, x_1^{2*}; x_2^{1*}, x_2^{2*}) = \left(2\alpha, \frac{2\alpha}{1+\alpha\beta}; 2(1-\alpha), \frac{2(1-\alpha+\alpha\beta)}{1+\alpha\beta}\right)$. If the asserted utility comparison holds for $t = 0$ and some sufficiently large γ, then by continuity, it still holds for small positive t and the same value of γ.

We observe that the overall commodity allocation is Pareto inefficient. We also note that households with efficient net trades are not affected by inefficient demand choices of other households.

References

Baumol, W. I., Panzar, J. C., & Willig, R. D. (1982). *Contestable markets and the theory of industry structure*. New York, NY: Harcourt Brace Jovanovich.

Becker, G. S. (1978). *The economic approach to human behavior*. Chicago, IL: The University of Chicago Press.

Becker, G. S. (1981). *A treatise on the family*. Cambridge, MA: Harvard University Press.

Bergstrom, T. C. (1989). A fresh look at the rotten kid theorem–and other household mysteries. *Journal of Political Economy, 97*(5), 1138–1159.

Browning, M., Chiappori, P. -A., & Lechene, V. (2010). Distributional effects in household models: Separate spheres and income pooling.*The Economic Journal, 120*, 786–799.

Chiappori, P.-A. (1992). Collective labor supply and welfare. *Journal of Political Economy, 100*, 437–467.

Gersbach, H., & Haller, H. (1999). Allocation among multi-member households: Issues, cores and equilibria. In A. Alkan, C. D. Aliprantis, & N. C. Yannelis (Eds.), *Current trends in economics: Theory and applications*. Berlin, Heidelberg: Springer-Verlag.

Gersbach, H., & Haller, H. (2001). Collective decisions and competitive markets. *Review of Economic Studies, 68*, 347–368.

Gersbach, H., & Haller, H. (2005). When inefficiency begets efficiency. *Economic Theory, 25*, 105–121.

Gersbach, H., & Haller, H. (2006). Household inefficiency and equilibrium efficiency. In C. Schultz & K. Vind (Eds.), *Institutions, equilibria and efficiency: Essays in honor of Birgit Grodal* (pp. 187–209). Heidelberg: Springer-Verlag.

Haller, H. (2000). Household decisions and equilibrium efficiency. *International Economic Review, 41*, 835–847.

Hart, O. D. (1983). The market mechanism as an incentive scheme. *Bell Journal of Economics,, 14*, 366–382.

Lechene, V., & Preston, I. (2011). Noncooperative household demand. *Journal of Economic Theory, 146*, 504–527.

Sato, N. (2009). Existence of competitive equilibrium in economies with multi-member households. *Economics Bulletin, 29*, 1760–1771.

Schumpeter, J. A. (1975). *Capitalism, socialism and democracy*. New York, NY: Harper.

Chapter 14
Public Goods and Public Choice

In this chapter, we demonstrate how general equilibrium models with multi-member households can be combined with public choice. Public choice refers to collective decisions of the entire society regarding global public good provision, financing and redistribution. Global public goods benefit all individuals in the economy.[1] Such an extension introduces a variety of conceptual issues as global collective decisions regarding public goods and public financing have to be integrated into the general equilibrium framework of Gersbach and Haller (2011). In the sequel, we are going to outline a flexible approach that allows to extend the equilibrium notions in the previous chapters by *public choice*. A more detailed investigation with several new results is carried out in Gersbach and Haller (2014).

14.1 Consumer Characteristics and Allocations

In this section, we present the basic structure of the model. It consists of decision units (consumers, households, and the government), the objects of decisions (private and public goods, households), decision criteria (preferences), and outcomes (allocations). Essentially, we integrate public good provision, global collective decisions and (in a very rudimentary form) the state into the general equilibrium model of Gersbach and Haller (2011) which is described in Chap. 7.

Consumers and Household Structures. We consider a finite population of **consumers**, represented by a set $I = \{1, \ldots, N\}$. A generic consumer is denoted i or j. The population I is partitioned into households. That is, there exists a partition P

[1]Public goods are non-rivalrous in consumption and non-excludable. The equilibrium notion can be readily extended to excludable public goods.

H. Gersbach and H. Haller, *Groups and Markets*,
DOI 10.1007/978-3-319-60516-6_14

of I into non-empty subsets referred to as households. For a consumer $i \in I$, $P(i)$ denotes the unique element of P (unique household in P) to which i belongs. If a partition P consists of H households, we frequently label them $h = 1, \ldots, H$, provided that this causes no confusion.

A (potential) household or group of consumers is any non-empty subset h of the population I. A generic household is denoted h. $\mathcal{H} = \{h \subseteq I | h \neq \emptyset\}$ denotes the set of all potential households. For $i \in I$, $\mathcal{H}_i = \{h \subseteq I \,|\, i \in h\}$ denotes the set of all potential households which have i as a member.

We call any partition P of I a **household structure in** I. We treat the household structure as an object of endogenous choice: Households are formed so that some household structure P is ultimately realized. Consequently, our **consumer allocation space** is $\mathcal{P}$, the set of all household structures in I.

Relative to a household structure P, we use the following terminology regarding $i \in I$ and $h \subseteq I$, $h \neq \emptyset$:

"household h exists" or "household h is formed"		iff $h \in P$;
"i belongs to h" or "individual i is a member of household h"		iff $i \in h$.

Private Commodities. There exists a finite number $\ell \geq 1$ of private commodities. Thus the private commodity space is $\mathbb{R}^{\ell}$. Private commodities are denoted by superscripts $k = 1, \ldots, \ell$. Each private commodity is formally treated as a private good, possibly with externalities in consumption. That is, private commodities are rival in consumption and a particular household's ownership excludes other households from consumption. Consumer $i \in I$ has private consumption set $X_i = \mathbb{R}^{\ell}_{+}$ so that the **private commodity allocation space** is $\mathcal{X} \equiv \prod_{j \in I} X_j$. Generic elements of $\mathcal{X}$ are denoted $\mathbf{x} = (x_i)$, $\mathbf{y} = (y_i)$ with $x_i = (x_i^1, \ldots, x_i^{\ell})$, $y_i = (y_i^1, \ldots, y_i^{\ell})$. For a potential household $h \subseteq I$, $h \neq \emptyset$, we set $\mathcal{X}_h = \prod_{i \in h} X_i$, the consumption set for household h. $\mathcal{X}_h$ has generic elements $\mathbf{x_h} = (x_i)_{i \in h}$. If $\mathbf{x} = (x_i)_{i \in I} \in \mathcal{X}$ is a private commodity allocation, then consumption for household h is the restriction of $\mathbf{x} = (x_i)_{i \in I}$ to h, $\mathbf{x_h} = (x_i)_{i \in h}$.

Endowments with Private Commodities. For a potential household $h \subseteq I$, $h \neq \emptyset$, its **endowment** is a private commodity bundle $\omega_h \in \mathbb{R}^{\ell}$ given by the sum of the endowments of all participating individuals: $\omega_h = \sum_{i \in h} \omega_{\{i\}}$ where $\omega_{\{i\}}$ is the endowment when individual i forms a single-person household. The **social endowment with private commodities** is given as

$$\omega_S \equiv \sum_{h \in P} \omega_h = \sum_{i \in I} \omega_{\{i\}}. \tag{14.1}$$

Note that the social endowment is independent of the household structure.

Public Goods. There exists a finite number $q \geq 1$ of **public goods**. Those goods are non-excludable and non-rivalrous in consumption. The public good space is $\mathbb{R}^q$. The consumption set for public goods is $\mathbb{R}^q_+$. A generic bundle of public goods is denoted by $g = (g_1, \ldots, g_q) \in \mathbb{R}^q_+$ where for $m = 1, \ldots, q$, g_m is the amount of public good m.

Provision of Public Goods. Public goods are produced by the government (or on behalf of the government) with the use of private commodities. In particular,

$$g_m \leq F_m(v_m)$$

where $v_m = (v_m^1, \ldots, v_m^\ell) \in \mathbb{R}^\ell_+$ is the vector of private commodities used to produce the amount g_m of the public good m. The production functions $F_m : \mathbb{R}^\ell_+ \to \mathbb{R}_+$ $(m = 1, \ldots, q)$ are assumed to be continuous and concave.

Allocations. An **allocation** is a triple $(\mathbf{x}, g; P) \in \mathcal{X} \times \mathbb{R}^q_+ \times \mathcal{P}$ specifying an allocation bundle of private commodities, a bundle of public goods, and household membership of each consumer. We call an allocation $(\mathbf{x}, g; P) \in \mathcal{X} \times \mathbb{R}^q_+ \times \mathcal{P}$ **feasible** if there exist $v_m \in \mathbb{R}^\ell_+$ $(m = 1, \ldots, q)$ such that

$$\sum_{i \in I} x_i + \sum_{m=1}^{q} v_m = \omega_S; \tag{14.2}$$

$$g_m = F_m(v_m) \quad \text{for} \quad m = 1, \ldots, q. \tag{14.3}$$

After the specification of individual preferences, by means of utility representations, an allocation determines the welfare of each and every member of society. In the current model, feasibility of an allocation does not depend on the household structure while its desirability may well depend on it.

Consumer Preferences. In principle, a consumer might have preferences on the allocation space $\mathcal{X} \times \mathbb{R}^q_+ \times \mathcal{P}$ and care about each and every detail of an allocation. But we shall restrict our analysis to situations of **household-specific preferences**, expanding the corresponding notion in Gersbach and Haller (2011) to economies with public goods. Consumers with such preferences care about public goods. They do not care about consumption of private commodities and household composition beyond the boundaries of their own household. That is, given a particular household structure, an individual is indifferent with respect to the affiliation and consumption of individuals belonging to other households.

We are going to make the **Assumption of Household-Specific Preferences (HSP)** throughout this chapter. We represent these preferences by utility functions. To this end, let us denote $\mathcal{X}^* = \bigcup_{h \in \mathcal{H}} \mathcal{X}_h$ and define $\mathcal{A}_i = \{(\mathbf{x_h}, g; h) \in \mathcal{X}^* \times \mathbb{R}^q_+ \times \mathcal{H} : h \in \mathcal{H}_i, \mathbf{x_h} \in \mathcal{X}_h, g \in \mathbb{R}^q_+\}$ for $i \in I$. We assume that each individual $i \in I$ has a UTILITY REPRESENTATION $U_i : \mathcal{A}_i \to \mathbb{R}$.

The assumption (HSP) has been extensively justified in Gersbach and Haller (2011). Among other things, it allows for **local public goods** within households. Private commodity k virtually constitutes a local public good for household h if for all members $i \in h$, U_i does not depend on individual consumption, but only on the aggregate consumption $\sum_{i \in h} x_i^k$ of good k by household members.

It is useful to distinguish several special cases of (HSP).

(PGE) Pure Group Externalities: For each consumer i, there exist functions $U_i^C : X_i \times \mathbb{R}_+^q \to \mathbb{R}$ and $U_i^G : \mathcal{H}_i \to \mathbb{R}$ such that
$U_i(\mathbf{x_h}, g; h) = U_i^C(x_i, g) + U_i^G(h)$ for $g \in \mathbb{R}_+^q, h \in \mathcal{H}_i, \mathbf{x_h} \in \mathcal{X}_h$.

(SEP) Separable Preferences: For each consumer i, there exist functions $U_i^c : X_i \to \mathbb{R}$, $V_i^c : \mathbb{R}_+^q \to \mathbb{R}$ and $U_i^G : \mathcal{H}_i \to \mathbb{R}$ such that
$U_i(\mathbf{x_h}, g; h) = U_i^c(x_i) + V_i^c(g) + U_i^G(h)$ for $g \in \mathbb{R}_+^q, h \in \mathcal{H}_i, \mathbf{x_h} \in \mathcal{X}_h$.

14.2 The Equilibrium Concept

In order to formulate an equilibrium of the economic system, several aspects have to be considered. At the conceptual level, we have to integrate household formation, collective decisions by households, market clearing and collective decisions on public good provision by the entire polity. In particular, we have to combine local collective decisions at the household level and global collective decisions on public good provision. There are several ways to formulate the latter. Global collective decisions involve decisions on the bundle of public goods the society is offering to its members and on how the expenditures are financed. Here we present a simple variant which requires that (a) at the going market prices and at the given household structure, a proposed bundle of public goods, g, cannot be improved by a coalition of households which together comprise a set of individuals of size n $(1 \leq n \leq N)$ and that (b) public good provision is financed by linear income taxes at the household level. The case $n = (N + 1)/2$ when N is uneven corresponds to the requirement that g is a Condorcet winner. For $n = N$, the requirement says that a change of the bundle of public goods has to be a Pareto improvement.

For the definition of an equilibrium notion, we first introduce a series of potential conditions:

- a household chooses collectively an efficient consumption schedule for its members, subject to the household disposable income;
- markets clear;
- no individual has an incentive to leave a household and to participate as an individual in the market at the going prices and at the current tax rate;
- no individual can leave a household and get accepted by another household by proposing a feasible allocation for the enlarged household which makes everybody in this newly formed household better off at the going prices and at the current tax rate;

- no group of individuals can benefit from forming a new household at the going prices and at the current tax rate;
- the budget of the public sector is balanced;
- no subset of households which comprises at least n individuals can propose an alternative bundle of public goods and a corresponding tax rate that will balance the public budget and make all individuals in these households weakly better off and at least one individual strictly better off at the going market prices.

The combination of all or a particular subset of the conditions defines a particular equilibrium notion. We next provide formal definitions of all conceivable equilibrium notions.

14.2.1 Definitions

We fix n at some level and we consider a household $h \in P$ and a price system $p \in \mathbb{R}^{\ell}$. For $\mathbf{x_h} = (x_i)_{i \in h} \in \mathcal{X}_h$,

$$p * \mathbf{x_h} \equiv p \cdot \left(\sum_{i \in h} x_i \right)$$

denotes the expenditure of household h on household consumption plan $\mathbf{x_h}$ at the price system p. As p and $\mathbf{x_h}$ are of different dimension for multi-member households, we use the $*$-product in lieu of the familiar inner product. Then h's **budget set** is defined as

$$B_h(p, t) = \{\mathbf{x_h} \in \mathcal{X}_h \ : \ p * \mathbf{x_h} \leq (1 - t) p \cdot \omega_h\}$$

where t is the tax rate on the nominal value of the endowments. We next define the **efficient budget set** $EB_h(p, t)$ as the set of $\mathbf{x_h} \in B_h(p, t)$ with the property that there is no $\mathbf{y_h} \in B_h(p, t)$ such that

$$U_i(\mathbf{y_h}, g; h) \geq U_i(\mathbf{x_h}, g; h) \quad \text{for all } i \in h;$$
$$U_i(\mathbf{y_h}, g; h) > U_i(\mathbf{x_h}, g; h) \quad \text{for some } i \in h.$$

Further define a **state** of the economy as a tuple $(p, \mathbf{x}, g, t; P)$ such that $p \in \mathbb{R}^{\ell}$ is a price system, $t \in [0, 1]$ is a tax rate and $(\mathbf{x}, g; P) \in \mathcal{X} \times \mathbb{R}^q_+ \times \mathcal{P}$ is an allocation, i.e., $\mathbf{x} = (x_i)_{i \in I}$ is an allocation of private commodities, $g = (g_1, \ldots, g_q)$ a bundle of public goods and P is an allocation of consumers (a household structure, a partition of the population into households). We say that in state $(p, \mathbf{x}, g, t; P)$,

(a) consumer i can benefit from exit, if $P(i) \neq \{i\}$ and there exists $y_i \in B_{\{i\}}(p, t)$ such that $U_i(y_i, g; \{i\}) > U_i(\mathbf{x_{P(i)}}, g; P(i))$;
(b) consumer i can benefit from joining another household h', if $h' \in P$, $h' \neq P(i)$ and there exists $\mathbf{y_{h' \cup \{i\}}} \in B_{h' \cup \{i\}}(p, t)$ such that $U_j(\mathbf{y_{h' \cup \{i\}}}, g; h' \cup \{i\}) > U_j(\mathbf{x_{P(j)}}, g; P(j))$ for all $j \in h' \cup \{i\}$.

(c) a group of consumers h can benefit from forming a new household, if $h \notin P$ and there exists $\mathbf{y_h} \in B_h(p, t)$ such that $U_j(\mathbf{y_h}, g; h) > U_j(\mathbf{x_{P(j)}}, g; P(j))$ for all $j \in h$.

Definition. A state $(p, \mathbf{x}, g, t; P)$ is a **competitive equilibrium with fixed household structure** P if there exist $v_m \in \mathbb{R}^{\ell}_{+}$, $m = 1, \ldots, q$, and the following conditions hold:

1. $\mathbf{x_h} \in EB_h(p, t)$ for all $h \in P$.
2. $\sum_{i \in I} x_i + \sum_{m=1}^{q} v_m = \omega_S$.
3. $g_m = F_m(v_m)$ for $m = 1, \ldots, q$.
4. $\sum_{m=1}^{q} p v_m \leq t p \sum_{h \in P} \omega_h$.

A state $(p, \mathbf{x}, g, t; P)$ is a **competitive equilibrium with free exit (CEFE)** if it satisfies conditions 1, 2, 3, 4 and

5. No consumer can benefit from exit.

A state $(p, \mathbf{x}, g, t; P)$ is a **competitive equilibrium with free household formation** if it satisfies conditions 1, 2, 3, 4, 5 and

6. No consumer can benefit from joining another household.

A state $(p, \mathbf{x}, g, t; P)$ is a **competitive equilibrium with new household formation** if it satisfies 1, 2, 3, 4, 5, 6 and

7. No group of consumers can benefit from forming a new household.

Finally, we introduce the following condition:

8. There exists no coalition Ω of households with at least n individuals that can propose an alternative public good bundle and financing scheme (g', t') such that

$$\sum_{m=1}^{q} p v'_m \leq t' p \sum_{h \in P} \omega_h;$$
$$g'_m = F_m(v'_m) \quad \text{for} \quad m = 1, \ldots, q;$$

with

$$U_i(\mathbf{x'_h}, g'; h) \geq U_i(\mathbf{x_h}, g; h) \quad \text{for all } i \in h \text{ with } h \in \Omega;$$
$$U_i(\mathbf{x'_h}, g'; h) > U_i(\mathbf{x_h}, g; h) \quad \text{for some } i \in h \text{ with } h \in \Omega;$$
$$\mathbf{x'_h} \in B_h(p, t') \quad \text{for all } h \in \Omega.$$

We can now add condition 8 to the equilibrium notions defined above. The corresponding equilibrium notions are extended by *"and public choice"*. For instance, if conditions 1, 2, 3, 4, 5, 6 and 8 hold, we obtain a competitive equilibrium with free household formation *and public choice*.

14.2.2 *Alternative Condition for Public Choice*

The equilibrium with public choice assumes that households proposing an alternative scheme (g', t') recognize that an alternative bundle of public goods may require a different tax rate to balance the budget. However, the coalition Ω does not take into account that different public good bundles may impact on prices of private commodities as they continue to be price takers in the sphere of private consumption. One could formulate an equilibrium notion in which more demanding requirements on the coalition Ω are made. For this purpose we formulate the following condition:

8*. There exists no coalition Ω of households with at least n individuals that can propose a public good bundle and financing scheme (g', t') such that the state of the economy $(p', \mathbf{x}', g', t'; P)$ is a competitive equilibrium with free exit (or free household formation or new household formation) with household structure P and

$$U_i(\mathbf{x}'_{\mathbf{h}}, g'; h) \geq U_i(\mathbf{x}_{\mathbf{h}}, g; h) \quad \text{for all } i \in h \text{ with } h \in \Omega;$$
$$U_i(\mathbf{x}'_{\mathbf{h}}, g'; h) > U_i(\mathbf{x}_{\mathbf{h}}, g; h) \quad \text{for some } i \in h \text{ with } h \in \Omega;$$

As shown in Gersbach and Haller (2014), there are instances where condition 8 is satisfied while condition 8* is not. However, there are also instances where condition 8* is satisfied while condition 8 is not.

14.2.3 *Adding Redistribution*

The preceding equilibrium notions involve government expenditures solely for the purpose of financing public goods. The equilibrium notions can easily be extended to incorporate redistribution. The modified equilibrium conditions are as follows:

$\widehat{1}$. $\mathbf{x}_{\mathbf{h}} \in EB_h(p, t, s_h)$ for all $h \in P$,
where s_h is the nominal subsidy that household h obtains from the state and $EB_h(p, t, s_h)$ is the efficient budget set of household h when h's budget set is $B_h(p, t, s_h) = \{\mathbf{x}_{\mathbf{h}} \in \mathcal{X}_h \ : \ p * \mathbf{x}_{\mathbf{h}} \leq (1 - t)p \cdot \omega_h + s_h\}$.

$\widehat{4}$. $\sum_{m=1}^{q} pv_m + \sum_{h \in P} s_h \leq tp \sum_{h \in P} \omega_h$.

$\widehat{8}$. There exists no coalition Ω of households with at least n individuals that can propose an alternative public good bundle, a subsidy scheme $s' = \{s'_h\}_{h \in P}$ and financing scheme (g', t') such that

$$\sum_{m=1}^{q} pv'_m + \sum_{h \in P} s'_h \leq t' p \sum_{h \in P} \omega_h;$$
$$g'_m = F_m(v'_m) \quad \text{for } m = 1, \ldots, q;$$

with

$$U_i(\mathbf{x}'_\mathbf{h}, g'; h) \geq U_i(\mathbf{x}_\mathbf{h}, g; h) \quad \text{for all } i \in h \text{ with } h \in \Omega;$$
$$U_i(\mathbf{x}'_\mathbf{h}, g'; h) > U_i(\mathbf{x}_\mathbf{h}, g; h) \quad \text{for some } i \in h \text{ with } h \in \Omega;$$
$$\mathbf{x}'_\mathbf{h} \in B_h(p, t', s'_h) \quad \text{for all } h \in \Omega.$$

The condition $\widehat{8}$ involves global collective decisions regarding taxation, subsidization and public good provision. While we have assumed in $\widehat{8}$ that redistribution does not entail losses of resources, costs of redistribution schemes can easily be incorporated into the budget constraint of the state as well.

14.3 Applications

The preceding framework lends itself to a number of potential and quite different applications, some of which are pursued in Gersbach and Haller (2014), but most of them are unexplored. We provide four examples.

First, standard themes in general equilibrium theory such as existence and welfare properties are the usual starting point of such an inquiry. Since our framework is fairly general and encompassing, existence of equilibria will be the exception rather than the rule. Household formation, local collective decisions and competitive markets on their own are already plagued by non-existence problems. The same holds for global collective decisions. These problems are likely to be amplified in a framework that integrates four types of allocation mechanisms: household formation, competitive markets, local and global collective decisions.

Second, if conditions are favorable for the stability of particular household structures in the presence of competitive markets, it remains to be examined how stability is affected by the additional presence of global collective decisions. In turn, it is an open question whether stable global collective decisions in a society in which each voter is a single will be impaired by multi-member household formation or whether to the contrary, household formation contributes to the stability of political decisions.

Third, it is well-known that without constraints or redistribution costs, collective decisions on redistribution may lead to extreme forms of the tyranny of the majority.[2] In order to limit such attempts of forming coalitions to exploit the remaining agents, constitutional limits on the taxation and subsidization of individuals and households are required. With such constitutional rules individuals' endowments are protected to some degree and all individuals can have positive consumption. Since this theme is an enduring challenge for constitutional economists when they explore collective

[2]The tyranny of the majority refers to a situation in which a set of citizens may be successful in channeling a large amount of the income of a minority via a tax/subsidy scheme into their own pockets.

decisions among voters acting as singletons, it will not be less challenging in our framework. Household formation may, for instance, affect the ease with which particular coalitions can exploit minorities in the society. Coalitions that include only some but not all members of households are less effective in exploiting minorities as redistribution within households cannot be avoided. In turn, coalitions consisting of a subset of households with particular characteristics may be easier to form. Whether household formation limits exploitation of minorities and may strike a different balance between the tyranny of the majority and minority protection is yet to be investigated.

Fourth, other fundamental issues of how democracy functions can be addressed in suitable variants of our model. For instance, should the weight of a citizen's vote in global collective decision be different when he or she is a member of a multi-member household compared to the vote cast as a single? Moreover, should the relative voting weights depend on the nature of the issue upon which society decides collectively? Some of these issues have been articulated for households with children and it has been suggested that the weight of an adult's vote in such households might be increased (see, for example, Sanderson and Scherbov (2007)). However, the issue is much more general. Moreover, recent inventions of voting procedures (see e.g. Gersbach (2004)) allow for more flexibility regarding the size of a majority that is required for the approval of a proposition. How such schemes can be adapted to take into account household formation remains to be seen.

References

Gersbach, H. (2004). Dividing resources by flexible majority rules. *Social Choice and Welfare*, *23*, 295–308.

Gersbach, H., & Haller, H. (2011). Competitive markets, collective decisions and group formation. *Journal of Economic Theory*, *146*, 275–299.

Gersbach, H. & Haller, H. (2014). Households, markets and public choice. CESifo Working Paper 4947.

Sanderson, W. C., & Scherbov, S. (2007). A near electoral majority of pensioners: Prospects and policies. *Population and Development Review*, *33*, 543–554.

Chapter 15
Economic Policy Analysis and Implications

Many of the empirical studies of household behavior are motivated by economic policy questions, which often have no straightforward answers. A general equilibrium approach helps capture relevant aggregate effects and spill-overs across households and markets. While our primary interest is purely theoretical, a number of our results lend themselves to policy interpretations and conclusions. We outline several themes in this chapter.

(I) First, we show by example the limited scope for social planning. The second welfare theorem in Gersbach and Haller (2001) demonstrates that the social planner can decentralize a fully Pareto optimal allocation as a competitive equilibrium among households by rearranging households and by lump-sum redistributions of endowments. The social planner faces, however, the requirement that households conform with his intentions when making their collective decisions. Allocations may be Pareto inefficient when a household takes another collective decision than envisioned by the social planner, as the next example demonstrates.

Example 6 of Gersbach and Haller (2001). Let $\ell = 1$ and $I = \{1, 2, 3\}$. Consumers $i = 1, 2$ experience no externalities. Consumer 3 is subject to consumption externalities. Specifically, we assume for $i = 1, 2$ that there exists a differentiable function $u_i : \mathbb{R}_+ \rightarrow \mathbb{R}$ with $u_i' > 0$ such that if $i \in h, h \subseteq I$, and $\mathbf{x_h} = (x_j)_{j \in h} \in \mathcal{X}_h$, then $U_i(\mathbf{x_h}; h) = u_i(x_i)$.

Especially, we assume for $i = 3, 3 \in h \subseteq I$, and $\mathbf{x_h} = (x_j)_{j \in h} \in \mathcal{X}_h$:

$$
\begin{aligned}
U_3(\mathbf{x_h}; h) &= x_3, \text{ in case } h = \{3\};\\
U_3(\mathbf{x_h}; h) &= x_3 + v(x_1), \text{ in case } h = \{1, 3\};\\
U_3(\mathbf{x_h}; h) &= x_3 + v(x_2), \text{ in case } h = \{2, 3\};\\
U_3(\mathbf{x_h}; h) &= x_3 + v(x_1 + x_2), \text{ in case } h = I;\\
\text{where } v(t) &= t - t^2 \text{ for } t \in \mathbb{R}.
\end{aligned}
$$

H. Gersbach and H. Haller, *Groups and Markets*,
DOI 10.1007/978-3-319-60516-6_15

We observe that (BE) is satisfied no matter what the household structure is. We further assume that the social endowment is $\omega = 3$ regardless of the household structure.

Suppose that the social planner wants to decentralize the allocation $(\mathbf{x^1}; P^1)$ where $P^1 = \{\{1\}, \{2, 3\}\}$ and $\mathbf{x^1} = (0, 0.5, 2.5)$. Clearly, $(\mathbf{x^1}; P^1) \in \mathcal{M}^*$. Let us pick the desired household structure $P^1 = \{\{1\}, \{2, 3\}\}$ and let lump-sum redistributions of commodities result in zero endowment for household $\{1\}$ and an endowment with three units of the commodity for household $\{2, 3\}$. Then $\mathbf{x^1}$ is a P^1-equilibrium allocation.

Suppose now that household $\{2, 3\}$ reaches another efficient collective decision, $\mathbf{z_h} = (\sqrt{3}, 3 - \sqrt{3})$. The resulting P^1-equilibrium allocation $\mathbf{y^1} = (0, \sqrt{3}, 3 - \sqrt{3})$ combined with P^1 lacks full Pareto efficiency, since the third consumer suffers a negative consumption externality. $(\mathbf{y^1}; P^1)$ is dominated by $(\mathbf{y^1}; P^2)$ with $P^2 = \{\{2\}, \{1, 3\}\}$, because the third consumer avoids negative consumption externalities when forming a household with the first consumer. □□

If one wanted to adopt a stronger equilibrium notion, a consumer ought to know what resources are available to alternative households. The social planner would have to assign property rights also for the alternative household structures not targeted by him. Let us consider the following property rights assignment:

$$
\begin{aligned}
P^0 &= \{\{1\}, \{2\}, \{3\}\} : \omega_{\{1\}} = 3,\ \omega_{\{2\}} = \omega_{\{3\}} = 0. \\
P^1 &= \{\{1\}, \{2, 3\}\} \quad : \omega_{\{1\}} = 0,\ \omega_{\{2,3\}} = 3. \\
P^2 &= \{\{2\}, \{1, 3\}\} \quad : \omega_{\{2\}} = 3,\ \omega_{\{1,3\}} = 0. \\
P^3 &= \{\{3\}, \{1, 2\}\} \quad : \omega_{\{3\}} = 3,\ \omega_{\{1,2\}} = 0. \\
P^4 &= \{I\} \qquad\qquad\quad : \omega_I \ = 3.
\end{aligned}
$$

With these specifications, the example persists under the strongest equilibrium notion in our context, a competitive equilibrium with new household formation—and also as an equilibrium in the club model of Ellickson et al. (2001).

The crucial property is that a consumer loses his stake in the household endowment, if he walks away from household $\{2, 3\}$. Consumer 1 on the other hand, cannot gain from joining household $\{2, 3\}$ to form the bigger household $\{1, 2, 3\}$, since he comes empty-handed. Thus the example persists for certain property rights assignments. It remains an open question what the planner can achieve in general by assigning property rights and relying on the Ellickson et al. (2001) equilibrium notion.

The example demonstrates that the possibilities of the social planner to decentralize Pareto efficient allocations are restricted by the nature of collective decision making of households. The social planner can, however, decentralize any Pareto efficient allocation as a competitive equilibrium among households if he rearranges households and redistributes endowments directly to individuals and if collective decisions within households are individually rational. E.g., individual rationality is satisfied if individuals can use their endowments as status quo points in a cooperative bargaining process.

The limitations of decentralization among households could have important implications for the design of social transfer systems. Reliance on efficient choices among households need not guarantee a fully Pareto optimal allocation if the recipient of transfers is the household as an entity or one distinguished member of the household. Individualized transfers are necessary to ensure the implementation of socially preferred allocations. However, the policy maker may be frustrated in his efforts to design individualized transfers, if individuals cannot protect their endowments in the collective choice process within households. For instance, a transfer meant to improve nutrition for the entire household, may result in a substitution effect, say increased cigarette consumption by the head of household. Targeted transfers might only serve their purpose if household expenditure could be monitored.

(II) In Gersbach and Haller (2009) and in Sect. 11.5, we demonstrate by way of examples that a shift of bargaining power within a two-person household can have three different intra-household effects in general equilibrium: (a) Only one member is affected. (b) The two members are affected in opposite ways. (c) Both members are affected in the same way. We also consider inter-household spill-overs. This kind of comparative statics is potentially applicable to instances of sociological changes as well as legal changes that cause a shift of bargaining power within some or all households.

(III) Further public policy issues in a variety of areas might be addressed within the current framework. We provide three examples. The first area consists of policies that directly affect outside options. For instance, laws governing the right to divorce, child support and marital property upon divorce influence directly the attractiveness of exercising outside options. Suppose future research shows that, as a rule, granting more outside options to individuals promotes social efficiency. Then divorce-related property rights should not depend on the presence of other adults in the new households of the ex-spouses. However, one of our examples in Gersbach and Haller (2011) suggests that the joining option can destabilize households and eliminate superior equilibria. If these negative effects turn out to be the rule, then taking the joining option should possibly be discouraged, for instance by granting less generous property rights to those who exercise it.

The second area comprises policies that influence consumption externalities. For instance, taxes that can affect consumption externalities may also affect the stability of households. Consider a two-person household which is formed because of positive group externalities. But one non-smoking member suffers from negative consumption externalities, because the partner is smoking. A sufficiently large cigarette tax reduces smoking and, depending on the elasticity of cigarette demand, may increase or reduce consumption of the remaining goods in the household. In the former case, a cigarette tax can enhance the stability of the particular household—which might serve as an additional argument in support of such a "sin tax". In the latter case, the non-smoker might prefer to leave the household. Thus a heavy "sin tax" may also destabilize certain households.

The last category includes taxes and transfers to and from households and their members. Consider for instance the case of female labor supply. A sizeable fraction of women do not work outside of their home when living with partners, but presumably would go to work if they were single—unless they went on welfare instead.

Whether or not such corner solutions occur depends among other things on household decisions regarding care for children but very likely also on how a second household income is treated with respect to taxation and transfers. High marginal tax rates on second household incomes occur naturally when taxation is progressive, only total household income is taxed and tax codes do not distinguish between multi-person and single-person households. Such tax systems tend to promote the aforementioned corner solutions. They may also make the exit option excessively attractive, conceivably with undesirable consequences. Hence they can influence both the allocation of resources and the composition of households.

(IV) Demographic changes in aging societies of industrial countries pose considerable challenges for economic policy. A particularly pressing issue is how to take care of elderly persons who rely on outside help to some extent—from support when buying goods to healthcare-related services. While such assistance could be bought in the market place, it might also be provided within multi-member households—or possibly within extended households. In such households, support for elderly persons is provided continuously by the same persons, who are rewarded through possible externalities and redistribution within the household. If larger households can efficiently take care of elderly people in need, policy may want to create favorable conditions for the formation or stability of such households. Then, numerous specific policy questions arise. To ensure the stability of this kind of arrangement, governments may want to provide transfers to those households, for instance, partially compensating persons able and willing to provide the necessary services to the elderly. Or the government may foster building standards according to which buildings would have to be particularly suitable for the formation of such households.

Naturally, while governments may help latent externalities to materialize within multi-member households in which services to elderly persons are provided, such interventions will come at a cost. Those costs have to be traded off against possible benefits. Suitable versions of our general equilibrium models promise to shed new light on such trade-offs and the nature of optimal policies whose characteristics—when examined from an encompassing perspective—are largely unknown. "Optimal" means second-best if one takes into account that the need for services and the quality of services rendered within a household (or those provided by third parties as well) can be difficult to observe for the government.

References

Ellickson, B., Grodal, B., Scotchmer, S., & Zame, W. R. (2001). Clubs and the market: Large finite economies. *Journal of Economic Theory*, *101*, 40–77.

Gersbach, H., & Haller, H. (2001). Collective decisions and competitive markets. *Review of Economic Studies*, *68*, 347–368.

Gersbach, H., & Haller, H. (2009). Bargaining power and equilibrium consumption. *Social Choice and Welfare*, *33*, 665–690.

Gersbach, H., & Haller, H. (2011). Competitive markets, collective decisions and group formation. *Journal of Economic Theory*, *146*, 275–299.

MIX
Papier aus verantwortungsvollen Quellen
Paper from responsible sources
FSC® C105338

If you have any concerns about our products,
you can contact us on
ProductSafety@springernature.com

In case Publisher is established outside the EU,
the EU authorized representative is:
Springer Nature Customer Service Center GmbH
Europaplatz 3, 69115 Heidelberg, Germany

Printed by Libri Plureos GmbH
in Hamburg, Germany